Márton Kuslits

Analysis and Optimisation of a New Differential Steering Concept

Logos Verlag Berlin

λογος

Bibliographic information published by the Deutsche Nationalbibliothek

The Deutsche Nationalbibliothek lists this publication in the Deutsche
Nationalbibliografie; detailed bibliographic data are available
on the Internet at http://dnb.d-nb.de .

Zugl.: Diss., BTU Cottbus-Senftenberg, 2022

Logos Verlag Berlin, 2022

ISBN 978-3-8325-5578-8

Logos Verlag Berlin GmbH
Georg-Knorr-Str. 4, Geb. 10,
D-12681 Berlin
Germany

Tel.:   +49 (0)30 / 42 85 10 90
Fax: +49 (0)30 / 42 85 10 92
https://www.logos-verlag.de

Acknowledgements

The present thesis is the result of several years' work spent mainly on private research, nevertheless in cooperation with the Chair of Engineering Mechanics and Vehicle Dynamics at the Brandenburg University of Technology, Cottbus, Germany.

In the first place, I express my gratitude to my doctoral advisor Prof. Dieter Bestle for his trust, guidance and even criticism which changed the way I see research and engineering. His dedication and selfless help were pivotal in the success of my work. I also thank Prof. Andreas Daberkow for reviewing my thesis and Prof. Burghard Voß for chairing the examination board.

During the early phase of my research I was affiliated with Széchenyi István University, Győr, Hungary, from where I thank Prof. Zoltán Horváth for helping the progression of this work by fostering the cooperation with Prof. Bestle, and Prof. Zoltán Varga for the inspiring discussions related to the research topic.

I also thank all colleagues from the Chair of Engineering Mechanics and Vehicle Dynamics for their kindness and support, and especially for organising the yearly 'Kolloquium', where I enjoyed exchanging ideas. Finally, I thank my beloved Rozi for her enduring patience and backing.

Leatherhead, United Kingdom   Márton Kuslits
September 2022

Abstract

Analysis and Optimisation of a New Differential Steering Concept

Keywords: differential steering, multi-body system, in-wheel motors, steering control, steer-by-wire, multi-objective optimisation

The emergence of electric drives opens up new opportunities in vehicle design. For example, powerful in-wheel motors provide unprecedented flexibility in chassis design and are suitable for distributed drive solutions where the behaviour of the vehicle may be further improved by, e.g., torque vectoring. However, vehicles equipped with distributed drives imply new, non-trivial vehicle dynamics control problems.

This work aims at a new differential steering concept relying only on passive steering linkages where the necessary steering moment about the kingpins is generated by traction force differences produced by in-wheel motors. For the analysis of the proposed steering concept, a tailored multi-body system model is introduced with emphasis on the dynamics of the steering linkages. Since the proposed concept is a steer-by-wire one inherently, it requires a control system which is also discussed.

As we do not have well-established design rules for the proposed kind of vehicles, this work overcomes the missing experience and explores the general applicability of such a new steering concept by using multi-objective optimisation. For this purpose, various design objectives and constraints are defined with respect to the dynamic, steady-state and low-speed steering performance of the vehicle. Since mechanical and control parts are strongly coupled, their parameters are optimised simultaneously.

The resulting behaviour of the proposed steering concept is investigated by various simulation experiments demonstrating a comparable steering performance of the new steering concept as that of conventional passenger cars.

Kurzfassung

Analyse und Optimierung eines neuen Differenziallenkungkonzepts

Schlüsselwörter: Differenziallenkung, Mehrkörpersystem,
Radnabenmotor, Lenkungsregelung, Steer-by-Wire,
Mehrkriterienoptimierung

Das Aufkommen von Elektroantrieben eröffnet neue Möglichkeiten in der Fahrzeugentwicklung. Beispielweise erlauben leistungsstarke Radnabenantriebe eine unvergleichliche Flexibilität beim Chassisentwurf und sind auch geeignet für verteilte Antriebslösungen, mit denen das Fahrzeugverhalten z.B. durch Torque-Vectoring weiter verbessert werden kann. Fahrzeuge mit verteilten Antrieben implizieren jedoch neue, nicht-triviale Fahrdynamikregelungsprobleme.

Im Mittelpunkt dieser Arbeit steht ein neues Differenziallenkungskonzept, das ausschließlich aus passiven Lenkachsen besteht und bei denen das notwendige Lenkmoment durch Zugkraftdifferenzen von Radnabenmotoren erzeugt wird. Zur Analyse des vorgeschlagenen Lenkkonzepts wird ein maßgeschneidertes Mehrkörpermodell mit Schwerpunkt auf der Dynamik der Lenkachsen vorgestellt. Da das vorgeschlagene Konzept inhärent ein Steer-by-Wire Lenksystem enthält, erfordert es ein Regelungssystem, das ebenfalls diskutiert wird.

Für die vorgeschlagene Fahrzeugart existieren keine etablierten Konstruktionsregeln, weshalb diese Arbeit die fehlenden Erfahrungen überwindet und die allgemeine Anwendbarkeit des neuen Lenkungskonzepts durch die Verwendung einer Mehrkriterienoptimierung untersucht. Zu diesem Zweck werden verschiedene Optimierungskriterien und Nebenbedingungen sowohl für dynamische und stationäre Fahrzustände als auch für das Lenkverhalten beim Parkieren des Fahrzeugs definiert. Da Mechanik und Regelung des Fahrzeugs stark gekoppelt sind, werden deren Parameter gleichzeitig optimiert.

Das resultierende Verhalten des vorgeschlagenen Lenkkonzepts wird durch verschiedene Simulationen untersucht, die ein vergleichbares Lenkverhalten des neuen Lenkkonzepts wie bei konventionellen Fahrzeugen zeigen.

Contents

List of Symbols and Acronyms

Mathematical Symbols

$\dot{}$	first time derivative	T	transpose
$\ddot{}$	second time derivative	+	Moore–Penrose inverse
∂	partial derivative	$\times$	cross product
d	total derivative	$\odot$	Hadamard product (element-wise multiplication of two matrices)
δ, δ'	variation		
$\mathscr{L}$	Laplace transform		
diag	diagonal matrix	$\wedge$	logical conjunction
		$\vert_{*,\,*}$	function evaluation at $*$

Latin Letters

A	transfer function gain	C	shape factor (tyre)
A	index set of parameters	$\boldsymbol{C}$	output matrix
$\boldsymbol{A}$	system matrix	$\mathcal{C}$	set of best performance candidates
$\boldsymbol{A}_r$	reference model system matrix	C_L	controller transfer function
a	kingpin–centreline distance	C_{F_α}	cornering stiffness
$a, \boldsymbol{a}$	acceleration	$C_{F_{\alpha 0}}$	nominal cornering stiffness
$\bar{\boldsymbol{a}}$	local acceleration	C_{M_α}	self-aligning stiffness
B	stiffness factor (tyre)	C_y	lateral force shape factor
B	index set of removed parameters	$\boldsymbol{C}_y$	position constraint Jacobian
$\boldsymbol{B}$	input matrix	C_z	self-aligning torque shape factor
$\boldsymbol{B}_r$	reference model input matrix		
b	track rod length	$\boldsymbol{C}_z$	velocity constraint Jacobian

c	steering arm dimension	$\boldsymbol{h}$	design constraints
$c, \boldsymbol{c}$	position constraint	h_e	curb edge height
$c_{1,...,10}$	tyre-specific parameters	h_f, h_r	front/rear steering trapeze height
c_f, c_r	front/rear kingpin stiffness	I	inertial frame
c_w	radial tyre stiffness	I	integral gain
D	peak factor (tyre)	$\boldsymbol{I}$	identity matrix
d	kingpin damping	$\boldsymbol{I}$	inertia tensor
E	curvature factor (tyre)	J	LQR cost function
E_y	lateral force curvature factor	J_b	vehicle body's principal moments of inertia
E_z	self-aligning torque curvature factor	J_w	wheel's principal moment of inertia
e	exponential function	J_{w_r}	wheel's rotational inertia
$\boldsymbol{e}$	unit vector	K	body-fixed frames
e_β, e_ω	tracking error	$\boldsymbol{K}$	feedback gain matrix
$\boldsymbol{F}$	feedforward gain matrix	k	number of selected candidates for evaluation
F_B	tangential bore force		
$\boldsymbol{F}^{\mathrm{d}}$	dynamic collision force	$\boldsymbol{k}$	generalised Coriolis and centrifugal forces
F_x	traction force		
F_y	lateral tyre force	$\hat{k}_w$	wheel–road contact condition
F_{y_0}	nominal lateral force		
F_z	vertical tyre force	$\boldsymbol{L}_R$	Jacobian of rotations
F_{z_0}	nominal tyre load	$\boldsymbol{L}_T$	Jacobian of translations
f	number of position DoFs	$\boldsymbol{l}^a$	applied moments
f	frequency	l_c	contact patch length
$\hat{f}$	logarithmic frequency	l_f, l_r	front/rear axle distance to CoG
$\boldsymbol{f}^a$	applied forces		
$\boldsymbol{f}_v$	velocity relation function	l_t	track width
G	transfer function	$\boldsymbol{M}$	mass matrix
$\boldsymbol{G}$	transfer matrix	$\tilde{\boldsymbol{M}}$	intermediate matrix
$\tilde{G}$	approximate transfer function	$\mathcal{M}$	manual design
		$\boldsymbol{\mathcal{M}}_*$	inverted, derived and evaluated mass matrix
g	number of velocity DoFs		
g	gravitational acceleration	$\tilde{\boldsymbol{\mathcal{M}}}_*$	inverted, derived and evaluated $\tilde{\boldsymbol{M}}$
h	height of CoG		

M_B	bore torque	$\hat{q}$	logarithmic state weight
M^{d}	disturbance torque produced by $\boldsymbol{F}^{\mathrm{d}}$	$\hat{\boldsymbol{q}}$	resultant generalised forces
M_S	self-aligning torque	R	trajectory radius
M_{S_0}	nominal self-aligning torque	$\boldsymbol{R}$	input weight matrix
		r	patch element dimension
M_x	tyre tilting torque	r	input weight
M_y	rolling resistance torque	r	number of rank-sorted designs
M_z	road-normal tyre torque		
m	number of elements	$\boldsymbol{r}$	position vector
m	gross vehicle mass	$\hat{r}$	logarithmic input weight
m_b	vehicle body mass	$\boldsymbol{r}_*$	control reference vector
m_w	wheel mass	$\hat{\boldsymbol{r}}_*$	pre-amplified reference
N	number of iterations	r_D	dynamic rolling radius
n	number of elements	r_f, r_r	front/rear scrub radius
n_c	number of constraints	r_P	bore radius
n_p	number of design parameters	$\boldsymbol{S}$	rotation matrix
		$\boldsymbol{S}$	solution of the Riccati-equation
n_s	number of sample set elements		
		s	complex frequency
O	origin	s_B	bore slip
$\mathcal{O}$	optimised design	T	execution time
$\boldsymbol{o}$	design objectives	$\boldsymbol{T}$	state projection matrix
o_β, o_ω	overshoot	$\tilde{\boldsymbol{T}}$	extended state projection matrix
P	proportional gain		
$\mathcal{P}$	set of feasible designs	$\widetilde{\boldsymbol{\mathcal{T}}}$	partial inverse of the extended projection matrix
$\mathcal{P}^E$	set of evaluated designs		
p	number of bodies	T_β, T_ω	rise time
$\boldsymbol{p}$	design vector	t	time
$\hat{\boldsymbol{p}}$	reduced design vector	t_f, t_r	front/rear mechanical trail
$\boldsymbol{p}_c$	control parameter vector		
$\boldsymbol{p}_m$	vector of mechanical parameters	t_p	pneumatic trail
		$\boldsymbol{u}$	input vector
$\boldsymbol{Q}$	state weight matrix	$v, \boldsymbol{v}$	velocity
q	state weight	v_c	critical speed
$\boldsymbol{q}$	generalised applied forces	v_T	tangential speed

v_t	transition speed	x_w	longitudinal wheel position
$\boldsymbol{W}_h$	hidden layer weight matrix	x_{w_0}	initial longitudinal wheel position
w_c	contact patch width	y	y-coordinate/axis
$\boldsymbol{w}_o$	output weight vector	$\boldsymbol{y}$	generalised coordinates
$\boldsymbol{w}_{h_0}$	hidden layer bias vector	$\hat{\boldsymbol{y}}$	system output
w_{o_0}	output bias	y_s	vehicle CoG coordinate
x	x-coordinate/axis	z	z-coordinate/axis
$\boldsymbol{x}$	state vector	$\boldsymbol{z}$	generalised velocities
$\tilde{\boldsymbol{x}}$	pseudo-state vector	z_w	vertical wheel position
x_e	longitudinal curb position	z_{w_0}	initial vertical wheel position
x_s	vehicle CoG coordinate		

Greek Letters

α	tyre slip angle	$\hat{\xi}, \hat{\boldsymbol{\xi}}$	metamodel response
$\boldsymbol{\alpha}$	angular acceleration	ρ	correlation coefficient
$\overline{\boldsymbol{\alpha}}$	local angular acceleration	σ	real part of s
α_{eq}	equivalent tyre slip angle	τ	time constant
β	sideslip angle	τ_d	desired time constant
γ	camber angle	ϕ	patch element dimension
$\boldsymbol{\gamma}$	remaining term of the acceleration constraint	φ_n, φ_2	scaling factor (tyre)
		φ_w	wheel rotation angle
$\tilde{\boldsymbol{\gamma}}$	intermediate vector	φ_{w_0}	initial wheel rotation angle
γ_0	penalty term offset		
Δ	difference	χ	vector of a subset of design parameters
δ	steering angle		
δ_f, δ_r	front/rear steering angle	ψ	yaw angle
$\delta_f^\star$	reference steering angle	$\boldsymbol{\Omega}, \Omega$	wheel rolling velocity
$\boldsymbol{\zeta}$	equation of generalised accelerations	ω	yaw rate
		$\boldsymbol{\omega}$	angular velocity
$\boldsymbol{\lambda}$	Lagrange multipliers	$\widehat{\omega}$	angular frequency
μ	road friction coefficient	ω_n	road-normal turning velocity of the wheel
$\xi, \boldsymbol{\xi}$	full model response		

Recurrent Indices

$'$	quantity in frame K_0	i, j, k	running index, $i, j, k \in \mathbb{N}$
$''$	quantity in frame K_i, $i \neq 0$	l, u	lower/upper bound
$-$	quantity in Magic Formula frame	max	maximum or upper limit
		min	minimum or lower limit
(i)	i^{th} element of a vector or a labelled set	ref	reference
		w	quantities related to wheels
∞	steady state		

Acronyms

CoG	centre of gravity
DAE	differential-algebraic equation
DDAS	differential drive assisted steering
DoE	design of experiments
DoFs	degrees of freedom
EoM	equations of motion
EPS	electric power steering
LQR	linear quadratic regulator
NSGA-II	non-dominated sorting genetic algorithm, 2nd version
ODE	ordinary differential equation
oLHS	optimised Latin hypercube sampling
RMS	root mean square
RSM	response surface model

1 Introduction

New powertrain technologies, especially the emergence of electric drives and the increasing number of actuators in vehicles open new opportunities in layout and chassis design, as well as for vehicle dynamics control. Considering a state-of-the-art chassis system, it typically consists of an electromechanical power steering system (EPS) at least on the front axle and a stability control system implemented by independent brake actuators. In addition, active rear axle steering might be applied optionally. The dynamics of the vehicle may be further improved by torque vectoring, i.e., by applying different traction forces to each single wheel in a controlled manner as discussed by MOTOYAMA ET AL. (1993). Electric drives are particularly useful for this purpose as they are suitable for distributed drive solutions, and they can vary their output torque, and thus the traction forces, almost immediately upon request. A new, noticeable example of such a driveline solution is the application of powerful in-wheel motors as shown by PEROVIC (2012). It is assumed that in-wheel motors can deliver the performance of conventional drivelines while they provide unparalleled flexibility in chassis design and control.

Vehicles equipped with multiple steering actuators and distributed drives may be characterised as over-actuated, implying non-trivial control problems as mentioned by WANG AND WANG (2011), for example. Over-actuation means in this case that the vehicle has less degrees of freedom (DoFs) than the number of actuators (OPPENHEIMER ET AL., 2006). In general, we may characterise the spatial motion of a vehicle by at least six DoFs, while the overall number of DoFs might be even higher by taking the motion of wheels and suspension components into account. However, we may neglect many DoFs when focusing only on the instantaneous steering response of the vehicle, as displacements along certain DoFs are limited or not relevant. Following RIEKERT AND SCHUNCK (1940), it is typical for steering studies that we consider the vehicle as a system with only two velocity DoFs, assuming that the magnitude of velocity v is constant, while the DoFs are the sideslip angle β and the yaw rate ω, see Fig. 1.1.

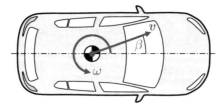

Figure 1.1: Velocity DoFs β and ω of a vehicle for $v = \text{const.}$

Electric vehicles with independently driven wheels and steering actuators are typical examples of over-actuated vehicles. Many studies utilise this property of electric vehicles, where the aim is usually to exploit the redundancy in order to set up a fail-safe system as demonstarted by WANG AND WANG (2011) or to find optimal control strategies as proposed by CHEN AND WANG (2012, 2014). Besides the obvious advantages of over-actuated vehicle systems, there are drawbacks, too. The higher number of actuators increases the system complexity and therefore the probability of faults, and considering an important aspect in automotive industry, the overall cost of the vehicle. Also, unwanted interference between the actuators may emerge (HARRER AND PFEFFER, 2017, pp. 547–548). In the recent years many researchers recognised that if the actuators are interchangeable in a vehicle, then some of them might be omitted. For example, once the traction is realised by independent drive capabilities, this property may be utilised for the steering of the vehicle. Under certain circumstances, the dedicated steering device(s) of the vehicle may then be simplified or even omitted.

An improved idea is the utilisation of the so-called *differential steering* principle. Differential steering is based on the difference of the drive torques (and hence traction forces) applied to the left and right wheels as sketched in Fig. 1.2 showing a MacPherson suspension with in-wheel motors and steering linkage. As traction forces generate torque about the kingpins for non-zero scrub radius of the suspension, a traction force imbalance $F_l \neq F_r$ results in a turn of the steering linkage during cornering, see Fig. 1.2b. During straight run, however, the traction forces are equal ($F_l = F_r$) cancelling out each others torque, see Fig. 1.2a. Moreover, the traction force difference may be superimposed with the traction forces applied for acceleration/deceleration of the vehicle, so the steering and acceleration/deceleration functions may be realised by the same actuators (drive motors) without significant perturbations in each others' operation. In other words, if $F_l + F_r \approx \text{const.}$ then the longitudinal dynamics of the vehicle is not influenced significantly during cornering.

(a)   (b)

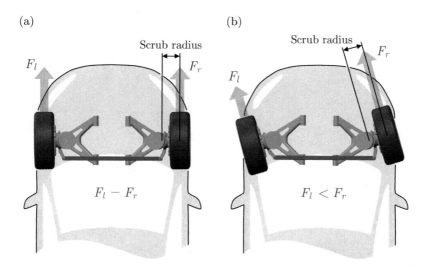

Figure 1.2: Basic principle of differential steering: (a) straight run and
(b) cornering

1.1 State of the Art in Differential Steering

Basically, differential steering means steering by the difference of longitudinal tyre forces between the left and right sides. In this sense, the simplest implementation of the differential steering principle is the so-called skid steering, widely used in heavy military and construction vehicles equipped with continuous tracks or pneumatic tyres., see Fig. 1.3 for example, or in commercial products like the Segway (KAMEN ET AL., 1994). In skid steering vehicles, the yaw moment is generated entirely by the longitudinal force difference produced either by differential braking or by differential driving realised by various mechanisms (McGUIGAN AND MOSS, 1998). However, skid steering may have drawbacks with respect to the cornering performance as presented by NAH ET AL. (2013) where the maximally reachable yaw rate is inferior compared to a conventionally steered vehicle in case of high-speed cornering manoeuvres. A further problem is that skid steering results in a high combined tyre slip which affects lateral stability of the vehicle adversely, increases the power consumption, and the wear of the tread and road surfaces (MACLAURIN, 2008).

An improved approach for vehicle steering is to exploit the *indirect* interaction between the traction forces and the self-steering properties of the suspension, in addition to the yaw moment produced by traction

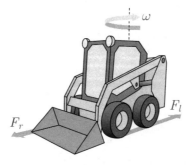

Figure 1.3: Skid-steered loader and its yaw motion

force differences. Typical examples of such vehicles are equipped with
swivel casters, Fig. 1.4. An early example is the electric three-wheeler of
Vedovelli and Priestley (DELASALLE, 1899). It features an active differential steering mechanism distributing the torque between its rear wheels,
and a swivel caster as the front wheel. A similar approach is applied in
mobile robots as well, see STAICU (2009) for example.

Figure 1.4: Swivel caster

Considering the *direct* interaction between longitudinal tyre forces and
suspension kinematics, the influence of traction and brake forces on
steering is traditionally deemed as an adverse effect and countered by
passive and active measures (DORNHEGE ET AL., 2017). The influence is
called either *torque steer* or *brake pull*, depending on whether it is produced
by accelerating or decelerating forces. It is particularly highlighted by
HARRER AND PFEFFER (2017, pp. 70–72) that the large scrub radius
largely contributes to the so-called torque steer effect. While the brake
pull is adverse in conventional steering systems, it provides the desired
steering effect in specific applications. Such application might be providing

emergency backup for steer-by-wire systems by wheel-individual braking as discussed by DOMINGUEZ-GARCIA ET AL. (2004) and GAUGER ET AL. (2016). JONASSON AND THOR (2018) introduced a model-based control method for this purpose and provided experimental verification as well.

Differential steering in the narrower sense utilises both traction and braking forces in a controlled manner for steering. WANG ET AL. (2008) introduced a vehicle model with four-wheel drive and differential steering on the front axle, and showed that the differential steering principle is suitable for steering assistance. Further, the same authors also introduced the term *differential drive assisted steering* (DDAS) from which the simplified and more general form *differential steering* is derived in the present work. WANG ET AL. (2009) extended the DDAS system with an additional direct yaw controller utilising the four-wheel drive layout, influencing not only the steering assistance but also the stability of the vehicle. The above mentioned results were later extended with advanced algorithms for vehicle speed estimation and tyre slip control, and the applicability of the DDAS system was also demonstrated in demanding manoeuvres like parking (WANG ET AL., 2011). A common property of these papers is the utilisation of the differential steering principle for steering assist where the authors additionally apply a conventional rack-and-pinion steering gear. It is also common that the rear axle is driven but not steerable. The steering assist controller and the direct yaw controller are designed and implemented separately and their cooperation is realised with an additional traction controller.

Many studies are focusing on particular control problems related to differential steering or applying advanced control methods. WU ET AL. (2013) investigated a two-wheel driven vehicle with a DDAS system focusing mainly on the control methods of the in-wheel electric motors from the steering point of view. Also this implementation still includes a rack-and-pinion steering mechanism and the rear axle is not utilised at all. LI ET AL. (2015) applied a fuzzy controller for the yaw response improvement of a vehicle with DDAS. ZHAO ET AL. (2013) presented a mixed sensitivity H_∞ method for robust control of the steering feel of a DDAS system taking into account the road disturbances and parameter uncertainties. ZHAO AND ZHANG (2018) later extended the previous paper with an additional yaw rate controller designed with μ-synthesis. The same group of authors also studied optimisation problems related to DDAS and its control (ZHAO ET AL., 2011, 2012, 2018, 2019).

A typical application of the differential steering principle is to use it as a safety fallback for conventional steer-by-wire systems implemented with dedicated steering actuators. Compared to the fallback solutions realised only by braking, differential steering provides an improved performance as it can actuate both traction and braking forces. For example, POLMANS AND STRACKE (2014) demonstrated such an application, while REITER ET AL. (2018) investigated the influence of the suspension kinematics on steering by comparing the performance of a conventional steer-by-wire system and a differential steering system as a fallback solution. KIRLI ET AL. (2017) discussed the limitations of differential steering as safety fallback and also proposed a control method to mitigate the emerged concerns. HU ET AL. (2019) also introduced a fallback solution for steer-by-wire systems by applying differential steering for lane keeping control realised with sliding mode methods. Similarly, TIAN ET AL. (2019) presented a solution by combining differential steering and direct yaw control.

As discussed above, the literature applies the differential steering principle typically for assistance or for safety fallback functions. Although WU ET AL. (2013), POLMANS AND STRACKE (2014) and REITER ET AL. (2018) imply the possibility of steering realised solely by the differential steering principle, the discussed literature always features a rack-and-pinion steering mechanism or a dedicated steering actuator. Besides steering assistance and steer-by-wire fallback, there are special aspects and use cases of differential steering systems as well. RÖMER ET AL. (2018) focused on energy consumption reduction, pointing out that a DDAS system reduces the energy demand compared to a conventional EPS while driven in the same driving cycle. ENGELMANN AND HERR (2018) applied differential steering to heavy mobile machines realising the solution uncommonly with a hydromechanical driveline powered by an internal combustion engine. WADEPHUL ET AL. (2018) developed a steering control method particularly for articulated vehicles.

1.2 Motivation and Outline of the Thesis

This work proposes a new differential steering concept and aims at the investigation of its fundamental behaviour and applicability. The concept is shown in Fig. 1.5, where the main novelties are that the differential steering principle is applied

- to both the front and rear axles by using four independently controlled in-wheel motors, and

- in a pure steer-by-wire manner completely avoiding any additional steering devices, unlike in the literature discussed above.

The work is motivated by the following potential advantages of the proposed concept:

- any kind of conventional steering system is omitted simplifying the vehicle significantly and thus reducing costs;
- the simplified chassis and the in-wheel motors provide more free space for the vehicle body and the layout design of the vehicle becomes more flexible;
- the new vehicle is *fully actuated* if one regards each axle as an actuator and compares this to the number of velocity DoFs in Fig. 1.1, i.e., there is no unnecessary actuator in the system and both velocity DoFs can be controlled independently which enables the realisation of any desired lateral dynamics characteristics.

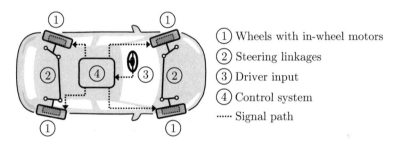

Figure 1.5: Proposed differential steering concept with four in-wheel motors and all-wheel steering

This work is organised in eight chapters. Following the introduction, Chapter 2 introduces a planar multi-body vehicle model of the concept shown in Fig. 1.5, where also tyre models are discussed. Chapter 3 discusses a novel symbolic linearisation method for non-holonomic systems with closed-loop kinematics and its application to the vehicle model. Subsequently, a control method derived from the linearised model is presented in Chapter 4. Chapter 5 features simulation studies and the formal characterisation of the steering performance. In Chapter 6, multi-objective optimisation of the steering performance is performed by utilising the performance characterisation. Chapter 7 investigates the disturbance rejection properties of the proposed steering concept. Finally, the work is closed by Chapter 8 with some conclusions and outlook. The above chapters are partially covered by studies already published by KUSLITS AND BESTLE (2018a,b, 2019, 2022).

2 Vehicle Model with Differential Steering

For developing the steering concept, we have to set up a vehicle model. There are various vehicle models in the literature from the linear single track model (RIEKERT AND SCHUNCK, 1940) to spatial multi-body models (BLUNDELL AND HARTY, 2015). To ease the modelling problem, one should choose the least complicated model that captures the relevant features subject to the intended investigation. Here, we have to consider that the differential steering principle strongly relies on dynamic interactions between the steering linkages and the vehicle body, while the dynamic interactions between the bodies are typically not considered in simple vehicle models.

As a compromise, a planar multi-body vehicle model is applied in this work, combining the simplicity of planar models with the ability of multi-body formalisms to capture dynamic interactions. Setting up such a model results in a non-holonomic problem with closed kinematic loops. Therefore, the equations of motion (EoM) are derived by using a combination of Jourdain's principle (JOURDAIN, 1909) and the virtual cut of the loops (see HWANG ET AL. (1990) for example) where the loop constraints are taken into account by the method of Lagrange multipliers derived from the constrained variation of the action functional of Hamilton's principle (HAND AND FINCH, 1998).

The chapter is organised as follows. The kinematics of the model is described first, followed by the derivation of the EoM. Subsequently, the applied tyre models of PACEJKA (2006) and HIRSCHBERG ET AL. (2007) are discussed.

2.1 Model Definition and Kinematics

Figure 2.1a shows the planar multi-body vehicle model which consists of five bodies, namely the vehicle body and four wheels attached to similar steering mechanisms at the front and rear axles. The vehicle has no conventional steering device, but steering is realised by the differential steering principle only. Four individually controlled in-wheel motors produce traction forces F_{x_i}, $i = 1 \ldots 4$, which result in steering torques about the kingpins if they are different between left and right sides. Lateral forces F_{y_i}, $i = 1 \ldots 4$, and torques M_{z_i}, $i = 1 \ldots 4$, also contribute passively to the resulting torques about the kingpins. The vehicle has both front and rear steering where, however, the concept may be reduced to front-axle steering as well.

The motion of the vehicle body is associated with the motion of the attached frame $K_0 \{O_0, x_0, y_0, z_0\}$ and may be described by the coordinates x_s and y_s of the centre of gravity (CoG) O_0 and the yaw angle ψ with respect to the inertial frame $I \{O_I, x_I, y_I, z_I\}$. The motion of the wheels may be characterised by the following geometrical descriptions. Front and rear axles are located at l_f and l_r distances from the CoG, respectively. Kingpins of the steering mechanisms have a distance a from the x_0-axis and b denotes the length of the track rods. In between the steering knuckles and the vehicle body, torsional springs and dampers characterised by stiffness coefficients c_f, c_r and damping coefficient d are connected. The geometry of the suspensions is defined in knuckle-fixed frames $K_i \{O_i, x_i, y_i, z_i\}$, $i = 1 \ldots 4$, where origins are located in the kingpins, see Fig. 2.1b and 2.1c. Mechanical trails of the front and rear axles are t_f and t_r, respectively, and r_f and r_r are the scrub radii. The steering arm in the knuckle-fixed frame is given by $c = a - b/2$ and h_f, h_r for front and rear axles, respectively. The orientations of the knuckle-fixed frames and thus the wheel orientations are described by steering angles δ_i, $i = 1 \ldots 4$.

For generating the EoM, we may cut the closed loops of the steering mechanisms by virtually removing the massless track rods, see Fig. 2.2. Later the track rods will be taken into account again by algebraic equations reducing the real position DoFs of the vehicle model to five. By applying the virtual cut, the motion of the resulting spanning tree may be uniquely described by the generalised coordinates

$$\boldsymbol{y} = \begin{bmatrix} x_s & y_s & \psi & \delta_1 & \delta_2 & \delta_3 & \delta_4 \end{bmatrix}^\mathrm{T} \in \mathbb{R}^f, \qquad (2.1)$$

where $f = 7$ is the number of position DoFs of the spanning tree system.

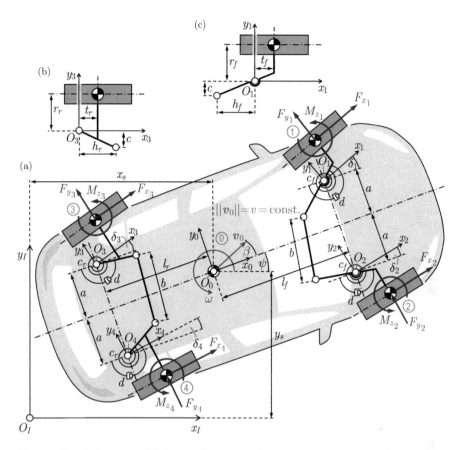

Figure 2.1: Planar vehicle model with front and rear wheel steering:
(a) main model and wheels with (b) rear and (c) front knuckle-
fixed frames

Analogously to RIEKERT AND SCHUNCK (1940), we may assume a con-
stant velocity $\|\boldsymbol{v}_0\| = v = \text{const.}$ of the CoG of the vehicle body, which
may be chosen arbitrarily. The constant velocity imposes a non-holonomic
constraint $\dot{x}_s = v \cos(\beta + \psi)$ and $\dot{y}_s = v \sin(\beta + \psi)$ on the velocity state
of the car body explicitly described by the sideslip angle β and yaw rate
$\omega = \dot{\psi}$, see Fig. 2.1a. Together with the steering velocities $\dot{\delta}_i$, this defines
the vector of generalised velocities

$$\boldsymbol{z} = \begin{bmatrix} \beta & \omega & \dot{\delta}_1 & \dot{\delta}_2 & \dot{\delta}_3 & \dot{\delta}_4 \end{bmatrix}^{\mathrm{T}} \in \mathbb{R}^g, \qquad (2.2)$$

where $g = 6$ is the number of velocity DoFs. Eventually, Fig. 2.1a and a

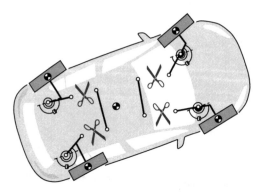

Figure 2.2: Virtual cut of track rods

comparison of $\boldsymbol{y}$ and $\boldsymbol{z}$ reveals the kinematic relation

$$\dot{\boldsymbol{y}} = \boldsymbol{f}_v(\boldsymbol{y}, \boldsymbol{z}) = \begin{bmatrix} v\cos(\beta + \psi) & v\sin(\beta + \psi) & \omega & \dot{\delta}_1 & \dot{\delta}_2 & \dot{\delta}_3 & \dot{\delta}_4 \end{bmatrix}^{\mathrm{T}}. \tag{2.3}$$

The motion of the vehicle body is given by position vector and rotation matrix

$$\boldsymbol{r}_{I0} = \begin{bmatrix} x_s \\ y_s \\ 0 \end{bmatrix}, \quad \boldsymbol{S}_{I0} = \begin{bmatrix} \cos\psi & -\sin\psi & 0 \\ \sin\psi & \cos\psi & 0 \\ 0 & 0 & 1 \end{bmatrix}, \tag{2.4}$$

and for the wheels the relative positions of the frame origins O_i with respect to K_0 are

$$\boldsymbol{r}'_{01} = \begin{bmatrix} l_f \\ a \\ 0 \end{bmatrix}, \quad \boldsymbol{r}'_{02} = \begin{bmatrix} l_f \\ -a \\ 0 \end{bmatrix}, \quad \boldsymbol{r}'_{03} = \begin{bmatrix} -l_r \\ a \\ 0 \end{bmatrix}, \quad \boldsymbol{r}'_{04} = \begin{bmatrix} -l_r \\ -a \\ 0 \end{bmatrix}. \tag{2.5}$$

The relative orientations of K_i follow as

$$\boldsymbol{S}_{0i} = \begin{bmatrix} \cos\delta_i & -\sin\delta_i & 0 \\ \sin\delta_i & \cos\delta_i & 0 \\ 0 & 0 & 1 \end{bmatrix}. \tag{2.6}$$

According to Figs. 2.1b and 2.1c, the wheel centre positions in K_i are

$$\boldsymbol{r}''_{11} = \begin{bmatrix} t_f \\ r_f \\ 0 \end{bmatrix}, \quad \boldsymbol{r}''_{22} = \begin{bmatrix} t_f \\ -r_f \\ 0 \end{bmatrix}, \quad \boldsymbol{r}''_{33} = \begin{bmatrix} t_r \\ r_r \\ 0 \end{bmatrix}, \quad \boldsymbol{r}''_{44} = \begin{bmatrix} t_r \\ -r_r \\ 0 \end{bmatrix}. \tag{2.7}$$

The absolute CoG positions and rotation matrices w.r.t. the inertial frame can then be determined as

$$r_i = \begin{cases} r_{I0} & \text{for } i = 0, \\ r_{I0} + S_{I0}\left(r'_{0i} + S_{0i}r''_{ii}\right) & \text{for } i = 1 \ldots 4 \end{cases} \tag{2.8}$$

and

$$S_i = \begin{cases} S_{I0} & \text{for } i = 0, \\ S_{I0}S_{0i} & \text{for } i = 1 \ldots 4. \end{cases} \tag{2.9}$$

Since the original kinematic loops have been removed for the above kinematic description, it is necessary to account for the track rods by $n_c = 2$ closing conditions derived in the body-fixed frame as

$$c = \begin{bmatrix} \left\|r'_{01} + S_{01}r''_{1_S} - \left(r'_{02} + S_{02}r''_{2_S}\right)\right\|^2 - b^2 \\ \left\|r'_{03} + S_{03}r''_{3_S} - \left(r'_{04} + S_{04}r''_{4_S}\right)\right\|^2 - b^2 \end{bmatrix} = 0, \tag{2.10}$$

where

$$r''_{1_S} = \begin{bmatrix} -h_f \\ -c \\ 0 \end{bmatrix}, \ r''_{2_S} = \begin{bmatrix} -h_f \\ c \\ 0 \end{bmatrix}, \ r''_{3_S} = \begin{bmatrix} h_r \\ -c \\ 0 \end{bmatrix}, \ r''_{4_S} = \begin{bmatrix} h_r \\ c \\ 0 \end{bmatrix}. \tag{2.11}$$

These closing conditions are discussed in more detail in Section 3.4.

The absolute velocities of the bodies' CoGs may be obtained from positions (2.8) by time differentiation as

$$v_i = \begin{cases} \dot{r}_{I0} & \text{for } i = 0, \\ \dot{r}_{I0} + \dot{S}_{I0}(r'_{0i} + S_{0i}r''_{ii}) + S_{I0}\dot{S}_{0i}r''_{ii} & \text{for } i = 1 \ldots 4. \end{cases} \tag{2.12}$$

Considering the orthogonality of the rotation matrices resulting in identities $S^{\mathrm{T}}S = I$, and the relation $\omega \times r = \dot{S}S^{\mathrm{T}}r$ for any vector r, we may substitute $\dot{S}r = \dot{S}S^{\mathrm{T}}Sr = \omega \times Sr$ in Eq. (2.12) and rewrite it as

$$v_i = \begin{cases} \dot{r}_{I0} & \text{for } i = 0, \\ \dot{r}_{I0} + \omega_{I0} \times S_{I0}(r'_{0i} + S_{0i}r''_{ii}) + S_{I0}(\omega_{0i} \times S_{0i}r''_{ii}) & \text{for } i = 1 \ldots 4, \end{cases} \tag{2.13}$$

where $\omega_{I0} = we_z$ and $\omega_{0i} = \dot{\delta}_i e_z$ are the angular velocities of K_0 and K_i relative to frames I and K_0, respectively. Alternatively, the absolute velocities may also be expressed as functions of generalised coordinates and velocities:

$$v_i(y, z) = \frac{\partial r_i}{\partial y}\dot{y} \equiv \frac{\partial r_i}{\partial y}f_v, \tag{2.14}$$

where $\dot{\boldsymbol{y}}$ is substituted according to Eq. (2.3).

The absolute accelerations follow from a further time differentiation of Eq. (2.13) as

$$
\boldsymbol{a}_i = \begin{cases}
\ddot{\boldsymbol{r}}_{I0} & \text{for } i = 0, \\[4pt]
\ddot{\boldsymbol{r}}_{I0} + \dot{\boldsymbol{\omega}}_{I0} \times \boldsymbol{S}_{I0}(\boldsymbol{r}'_{0i} + \boldsymbol{S}_{0i}\boldsymbol{r}''_{ii}) & \text{for } i = 1\ldots 4. \\[2pt]
+ \boldsymbol{\omega}_{I0} \times (\dot{\boldsymbol{S}}_{I0}\boldsymbol{r}'_{0i} + \dot{\boldsymbol{S}}_{I0}\boldsymbol{S}_{0i}\boldsymbol{r}''_{ii} + \boldsymbol{S}_{I0}\dot{\boldsymbol{S}}_{0i}\boldsymbol{r}''_{ii}) & \\[2pt]
+ \dot{\boldsymbol{S}}_{I0}(\boldsymbol{\omega}_{0i} \times \boldsymbol{S}_{0i}\boldsymbol{r}''_{ii}) & \\[2pt]
+ \boldsymbol{S}_{I0}(\dot{\boldsymbol{\omega}}_{0i} \times \boldsymbol{S}_{0i}\boldsymbol{r}''_{ii} + \boldsymbol{\omega}_{0i} \times \dot{\boldsymbol{S}}_{0i}\boldsymbol{r}''_{ii})
\end{cases}
$$

$$(2.15)$$

Similarly to Eq. (2.14), the absolute accelerations may also be expressed as functions of generalised coordinates, velocities and accelerations as

$$
\boldsymbol{a}_i(\boldsymbol{y}, \boldsymbol{z}, \dot{\boldsymbol{z}}) = \frac{\partial \boldsymbol{v}_i}{\partial \boldsymbol{z}}\dot{\boldsymbol{z}} + \frac{\partial \boldsymbol{v}_i}{\partial \boldsymbol{y}}\dot{\boldsymbol{y}} = \boldsymbol{L}_{T_i}(\boldsymbol{y}, \boldsymbol{z})\dot{\boldsymbol{z}} + \overline{\boldsymbol{a}}_i(\boldsymbol{y}, \boldsymbol{z}), \qquad (2.16)
$$

where $\boldsymbol{L}_{T_i} := \partial \boldsymbol{v}_i/\partial \boldsymbol{z}$ and $\overline{\boldsymbol{a}}_i := (\partial \boldsymbol{v}_i/\partial \boldsymbol{y})\,\boldsymbol{f}_v$ are the Jacobians and local acceleration terms (BESTLE, 1994, pp. 28–29).

The absolute angular velocities of the bodies may be directly deduced from Fig. 2.1 as

$$
\boldsymbol{\omega}_i = \begin{cases}
\dot{\psi}\boldsymbol{e}_z \equiv \omega\boldsymbol{e}_z & \text{for } i = 0, \\[4pt]
(\dot{\psi} + \dot{\delta}_i)\boldsymbol{e}_z \equiv (\omega + \dot{\delta}_i)\boldsymbol{e}_z & \text{for } i = 1\ldots 4.
\end{cases} \qquad (2.17)
$$

Similarly to Eq. (2.15), the absolute angular accelerations follow from the time derivation of the angular velocities (2.17) as

$$
\boldsymbol{\alpha}_i = \begin{cases}
\dot{\omega}\boldsymbol{e}_z & \text{for } i = 0, \\[4pt]
(\dot{\omega} + \ddot{\delta}_i)\boldsymbol{e}_z & \text{for } i = 1\ldots 4.
\end{cases} \qquad (2.18)
$$

Also the absolute angular accelerations may be written alternatively as

$$
\boldsymbol{\alpha}_i(\boldsymbol{z}, \dot{\boldsymbol{z}}) = \frac{\partial \boldsymbol{\omega}_i}{\partial \boldsymbol{z}}\dot{\boldsymbol{z}} + \frac{\partial \boldsymbol{\omega}_i}{\partial \boldsymbol{y}}\dot{\boldsymbol{y}} = \boldsymbol{L}_{R_i}\dot{\boldsymbol{z}} + \overline{\boldsymbol{\alpha}}_i \equiv \boldsymbol{L}_{R_i}\dot{\boldsymbol{z}}, \qquad (2.19)
$$

where $\boldsymbol{L}_{R_i} := \partial \boldsymbol{\omega}_i/\partial \boldsymbol{z}$ and $\overline{\boldsymbol{\alpha}}_i := (\partial \boldsymbol{\omega}_i/\partial \boldsymbol{y})\,\boldsymbol{f}_v \equiv \boldsymbol{0}$. Detailed results of the above derivations may be found in the Appendix.

2.2 Nonlinear Equations of Motion

Based on the kinematics, the EoM of the system with body masses m_i, inertia tensors I_i, and the applied forces and moments f_i^a, l_i^a, can be generated. For non-holonomic systems with $p = 5$ rigid bodies, they result from Jourdain's principle (BESTLE, 1994, p. 26)

$$\sum_{i=0}^{4} [\delta' v_i^{\mathrm{T}} (m_i a_i - f_i^a) + \delta' \omega_i^{\mathrm{T}} (I_i \alpha_i + \omega_i \times I_i \omega_i - l_i^a)] = 0, \qquad (2.20)$$

where $\delta' v_i = L_{T_i} \delta' z$ and $\delta' \omega_i = L_{R_i} \delta' z$ are variations of v_i and ω_i resulting from velocity variations $\delta' z$. Substitution yields

$$\delta' z^{\mathrm{T}} \sum_{i=0}^{4} [L_{T_i}^{\mathrm{T}} (m_i a_i - f_i^a) + L_{R_i}^{\mathrm{T}} (I_i \alpha_i + \omega_i \times I_i \omega_i - l_i^a)] = 0. \quad (2.21)$$

In general, the velocities z and thus variations $\delta' z$ are considered as independent. Here, however, the system is both constrained and non-holonomic requiring a special treatment. Starting from the position constraints (2.10), i.e., $c(y) = 0$ with no explicit time dependence, differentiation with respect to time yields

$$\dot{c} = \frac{\partial c}{\partial y} \dot{y} = C_y(y) f_v(y, z) = 0, \text{ where } C_y(y) := \frac{\partial c}{\partial y} \qquad (2.22)$$

and $\dot{y}$ is substituted according to Eq. (2.3). The variation of this velocity-restricting algebraic constraint with respect to z also enforces a constraint on velocity variations $\delta' z$ as

$$C_y \frac{\partial f_v}{\partial z} \delta' z =: C_z \delta' z = 0, \text{ where } C_z := C_y \frac{\partial f_v}{\partial z}. \qquad (2.23)$$

Embodying this in Eq. (2.21) by Lagrange multipliers $\lambda \in \mathbb{R}^2$ yields

$$\delta' z^{\mathrm{T}} \left\{ \sum_{i=0}^{4} [L_{T_i}^{\mathrm{T}} (m_i a_i - f_i^a) + L_{R_i}^{\mathrm{T}} (I_i \alpha_i + \omega_i \times I_i \omega_i - l_i^a)] - C_z^{\mathrm{T}} \lambda \right\}$$
$$= 0 \ \forall \, \delta' z,$$

$$(2.24)$$

where now $\delta' z$ may be considered as independent (BESTLE, 1994, p. 17). Thus, the variational equation is satisfied if

$$\sum_{i=0}^{4} [L_{T_i}^{\mathrm{T}} m_i a_i + L_{R_i}^{\mathrm{T}} (I_i \alpha_i + \omega_i \times I_i \omega_i)] - C_z^{\mathrm{T}} \lambda = \sum_{i=0}^{4} (L_{T_i}^{\mathrm{T}} f_i^a + L_{R_i}^{\mathrm{T}} l_i^a).$$

$$(2.25)$$

By substituting accelerations (2.16) and (2.19) in Eq. (2.25), the nonlinear EoM of the constrained non-holonomic system may be summarised as

$$M\left(y, z\right) \dot{z} + k\left(y, z\right) - C_z^{\mathrm{T}}\left(y, z\right) \lambda = q\left(y, z\right), \tag{2.26}$$

where

$$M = \sum_{i=0}^{4} \left(L_{T_i}^{\mathrm{T}} m_i L_{T_i} + L_{R_i}^{\mathrm{T}} I_i L_{R_i}\right), \tag{2.27}$$

$$k = \sum_{i=0}^{4} \left[L_{T_i}^{\mathrm{T}} m_i \overline{a}_i + L_{R_i}^{\mathrm{T}} I_i \overline{\alpha}_i + L_{R_i}^{\mathrm{T}} (\omega_i \times I_i \omega_i)\right], \tag{2.28}$$

$$q = \sum_{i=0}^{4} \left(L_{T_i}^{\mathrm{T}} f_i^a + L_{R_i}^{\mathrm{T}} l_i^a\right) \tag{2.29}$$

are the generalised mass matrix, vector of generalised Coriolis and centrifugal forces, and vector of generalised applied forces, respectively.

The body masses in Eqns. (2.27) and (2.28) are related to vehicle body $m_0 = m_b$ and wheels $m_i = m_w$, $i = 1 \ldots 4$, respectively. With rotation matrices (A.6), the inertia tensors with respect to the inertial frame follow as

$$
\begin{aligned}
I_0 &= S_{I0} \operatorname{diag}\left\{J_{b_x}, J_{b_y}, J_{b_z}\right\} S_{I0}^{\mathrm{T}} \\
&= \begin{bmatrix} J_{b_x} + \sin^2 \psi (J_{b_y} - J_{b_x}) & \sin\left(2\psi\right)\left(J_{b_x} - J_{b_y}\right)/2 & 0 \\ \sin\left(2\psi\right)\left(J_{b_x} - J_{b_y}\right)/2 & J_{b_y} + \sin^2 \psi (J_{b_x} - J_{b_y}) & 0 \\ 0 & 0 & J_{b_z} \end{bmatrix},
\end{aligned} \tag{2.30}
$$

with vehicle body's principal moments of inertia J_{b_x}, J_{b_y} and J_{b_z}, and

$$
\begin{aligned}
I_i &= S_i \operatorname{diag}\left\{J_w, J_{w_r}, J_w\right\} S_i^{\mathrm{T}}, \quad i = 1 \ldots 4 \\
&= \begin{bmatrix} J_w + \sin^2(\psi + \delta_i)(J_{w_r} - J_w) & \sin\left(2(\psi + \delta_i)\right)\left(J_w - J_{w_r}\right)/2 & 0 \\ \sin\left(2(\psi + \delta_i)\right)\left(J_w - J_{w_r}\right)/2 & J_{w_r} + \sin^2(\psi + \delta_i)(J_w - J_{w_r}) & 0 \\ 0 & 0 & J_w \end{bmatrix},
\end{aligned} \tag{2.31}
$$

with wheel's principal moments of inertia J_w about its transverse axes and J_{w_r} about its rotational axis, respectively.

The applied forces and torques in Eq. (2.29) may be deduced from Figure 2.1a with transformation matrices (A.6) as

$$
\boldsymbol{f}_i^a = \begin{cases}
\begin{bmatrix} 0 & 0 & -m_b g \end{bmatrix}^{\mathrm{T}} & \text{for } i = 0, \\[2mm]
\boldsymbol{S}_i \begin{bmatrix} F_{x_i} & F_{y_i} & F_{z_i} - m_w g \end{bmatrix}^{\mathrm{T}} & \text{for } i = 1 \ldots 4 \\[2mm]
= \begin{bmatrix} \cos(\psi + \delta_i) F_{x_i} - \sin(\psi + \delta_i) F_{y_i} \\ \sin(\psi + \delta_i) F_{x_i} + \cos(\psi + \delta_i) F_{y_i} \\ F_{z_i} - m_w g \end{bmatrix}
\end{cases}
\tag{2.32}
$$

summarising gravity force from gravitational acceleration g, longitudinal, lateral and vertical tyre forces F_{x_i}, F_{y_i} and F_{z_i}, respectively. Lateral forces result from the sideslip, whereas longitudinal forces might be produced almost arbitrarily by electric in-wheel motors. Vertical forces have no direct effect on the dynamics of the planar vehicle model and are listed here only for the sake of completeness. However, their indirect influence is discussed later in Section 2.3. Applied torques follow from road-normal torques M_{z_i} and from the kingpins' stiffness and damping as

$$
\boldsymbol{l}_i^a = \begin{cases}
\begin{bmatrix} 0 & 0 & c_f(\delta_1 + \delta_2) + c_r(\delta_3 + \delta_4) + d \sum_{j=1}^{4} \dot{\delta}_j \end{bmatrix}^{\mathrm{T}} & \text{for } i = 0, \\[3mm]
\begin{bmatrix} 0 & 0 & -c_f \delta_i - d\dot{\delta}_i + M_{z_i} \end{bmatrix}^{\mathrm{T}} & \text{for } i = 1, 2, \\[3mm]
\begin{bmatrix} 0 & 0 & -c_r \delta_i - d\dot{\delta}_i + M_{z_i} \end{bmatrix}^{\mathrm{T}} & \text{for } i = 3, 4.
\end{cases}
\tag{2.33}
$$

In order to simulate the vehicle behaviour, it is necessary to combine the EoM (2.26) with the second time-derivative of loop constraints (2.10) to differential-algebraic equations (DAE). The latter is identical with differentiating velocity constraints (2.22) with respect to time. Using the notation $\boldsymbol{\gamma} := [\partial(\boldsymbol{C_y f_v})/\partial \boldsymbol{y}]\,\boldsymbol{f}_v$ and substituting $\boldsymbol{C_z}$ according to Eq. (2.23) results in

$$
\ddot{\boldsymbol{c}} = \frac{\partial(\boldsymbol{C_y f_v})}{\partial \boldsymbol{y}}\boldsymbol{f}_v + \boldsymbol{C_y}\frac{\partial \boldsymbol{f}_v}{\partial \boldsymbol{z}}\dot{\boldsymbol{z}} \equiv \boldsymbol{C_z}\dot{\boldsymbol{z}} + \boldsymbol{\gamma} = \boldsymbol{0}.
\tag{2.34}
$$

The combination of Eqns. (2.26) and (2.34) results in the DAE

$$
\begin{bmatrix} \boldsymbol{M} & \boldsymbol{C_z}^{\mathrm{T}} \\ \boldsymbol{C_z} & \boldsymbol{0} \end{bmatrix}
\begin{bmatrix} \dot{\boldsymbol{z}} \\ -\boldsymbol{\lambda} \end{bmatrix} =
\begin{bmatrix} \boldsymbol{q} - \boldsymbol{k} \\ -\boldsymbol{\gamma} \end{bmatrix}.
\tag{2.35}
$$

Combined with Eq. (2.3), this builds up the state-space form

$$
\begin{bmatrix} \dot{y} \\ \dot{z} \\ -\lambda \end{bmatrix} = \left[\begin{bmatrix} M & C_z^T \\ C_z & 0 \end{bmatrix}^{-1} \begin{bmatrix} f_v \\ q-k \\ -\gamma \end{bmatrix} \right] \tag{2.36}
$$

of the vehicle model, where the Lagrange multipliers λ act as auxiliary variables. Detailed description of the EoM's components may be found in the Appendix. It should be noted that this is only a pseudo-state space representation due to internal dependencies in y and z. Some more discussion will be provided in Section 3.2 for the linear case.

2.3 Tyre Models

For the investigation of the differential steering concept, the choice of a proper tyre model is essential, as the tyre provides applied forces (2.32) and moments (2.33) to the vehicle model (2.36). For describing the tyre forces and the self-aligning torque, Pacejka's Magic Formula model is used (PACEJKA, 2006), while the bore torque is derived from a TMeasy-based model (HIRSCHBERG ET AL., 2007), where only the steady-state characteristics of the mentioned models are applied.

2.3.1 Modelling Considerations and Tyre Model Selection

The ISO 8855 standard defines all velocities, forces, torques and main dimensions related to the wheel, see Fig. 2.3, where v is the actual velocity of the wheel centre, Ω the rolling velocity of the wheel, α the sideslip angle, r_D the dynamic rolling radius of the wheel, and l_c the length of the contact patch. Regarding tyre forces and torques, F_x and F_y are the longitudinal and lateral forces as used above, F_z is the wheel load, M_x the tilting torque resulting from the wheel camber γ, M_y the rolling resistance torque, $M_z = M_S + M_B$ the sum of self-aligning torque M_S (see also Fig. 2.4) and bore torque M_B.

Since the vehicle model here is planar, it does not require a detailed suspension model. Further, the proposed steering concept does not depend significantly on the camber which is why the camber is neglected, i.e., $\gamma = 0$

resulting in $M_x = 0$. The rolling resistance is neglected or compensated by the in-wheel motors, respectively, resulting also in $M_y = 0$. It is supposed that the road surface is dry and, for the sake of simplicity, the friction ellipse is approximated by a friction circle; thus, the longitudinal and lateral friction coefficients are treated as equal values, where $\mu \approx 1$. The sideslip angle of each tyre follows from the wheel centre velocities $\boldsymbol{v}_i''$ as

$$\alpha_i = \arctan \left(\frac{v_{i_y}''}{v_{i_x}''} \right), i = 1 \dots 4, \tag{2.37}$$

where

$$\begin{aligned}
\boldsymbol{v}_i'' &= \boldsymbol{S}_i^{-1} \boldsymbol{v}_i = \boldsymbol{S}_i^{\mathrm{T}} \boldsymbol{v}_i = \boldsymbol{S}_{0i}^{\mathrm{T}} \boldsymbol{S}_{I0}^{\mathrm{T}} \boldsymbol{v}_i \\
&= \boldsymbol{S}_i^{\mathrm{T}} \dot{\boldsymbol{r}}_{I0} + \boldsymbol{S}_{0i}^{\mathrm{T}} \boldsymbol{S}_{I0}^{\mathrm{T}} [\boldsymbol{\omega}_{I0} \times \boldsymbol{S}_{I0}(\boldsymbol{r}_{0i}' + \boldsymbol{S}_{0i} \boldsymbol{r}_{ii}'')] + \boldsymbol{S}_{0i}^{\mathrm{T}} \boldsymbol{S}_{I0}^{\mathrm{T}} \boldsymbol{S}_{I0} (\boldsymbol{\omega}_{0i} \times \boldsymbol{S}_{0i} \boldsymbol{r}_{ii}'') \\
&= \boldsymbol{S}_i^{\mathrm{T}} \dot{\boldsymbol{r}}_{I0} + \boldsymbol{\omega}_{I0} \times (\boldsymbol{S}_{0i}^{\mathrm{T}} \boldsymbol{r}_{0i}' + \boldsymbol{r}_{ii}'') + \boldsymbol{\omega}_{0i} \times \boldsymbol{r}_{ii}''
\end{aligned} \tag{2.38}$$

is the transformation of absolute velocities (2.13) into wheel frame K_i with inverse rotation matrices (2.9), considering that $\boldsymbol{S}(\boldsymbol{\omega} \times \boldsymbol{r}) = \boldsymbol{S}\boldsymbol{\omega} \times \boldsymbol{S}\boldsymbol{r}$ for any rotation matrix $\boldsymbol{S}$ and $\boldsymbol{S}\boldsymbol{\omega} \equiv \boldsymbol{\omega}$ for planar models.

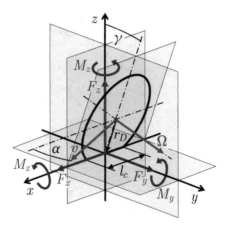

Figure 2.3: Wheel forces, torques, velocities and main dimensions according to ISO 8855

2.3.2 The Magic Formula Tyre Model

The basic Magic Formula

$$Y^- (\alpha) = D \sin \left(C \arctan \left(B\alpha - E \left(B\alpha - \arctan \left(B\alpha \right) \right) \right) \right) \tag{2.39}$$

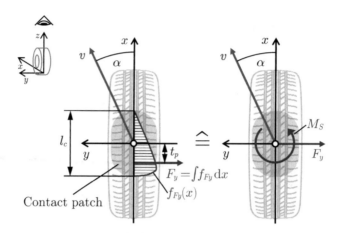

Figure 2.4: Pneumatic trail and generation of self-aligning torque M_S

is an empirical expression for calculating tyre forces and moments as function of sideslip angle α of the wheel, where B is a stiffness factor, C a shape factor, D the peak factor and E a curvature factor. The output variable Y^- either represents the nominal lateral force $F_{y_0}^-$ or the nominal self-aligning torque $M_{S_0}^-$ depending on how the factors are defined (PACEJKA, 2006, pp. 172–174). Note that the Magic Formula relies on a different reference frame than the one used in Fig. 2.3, which is why results of Eq. (2.39) have opposite sign compared to the ISO 8855 frame. This is distinguished by superscript '$-$'.

In general, the tyre characteristics may depend on various factors like the road friction coefficient μ, the camber angle γ and the vertical load F_z. For further calculations, however, first we have to determine the nominal tyre characteristics. Particularly, the nominal tyre characteristics result when the above mentioned influencing factors are nominal, i.e., $\mu \approx 1$, $\gamma = 0$ and $F_z = F_{z_0}$, where F_{z_0} is the nominal vertical load. As follows from Section 2.3.1, the influence of the road friction and the camber angle is neglected anyway, thus only F_z has to be taken into account.

Besides these considerations, the nominal lateral force $F_{y_0}^-(\alpha)$ results from Eq. (2.39) by substituting the stiffness factor as

$$B = \frac{C_{F_{\alpha_0}}}{CD},$$

(2.40)

where $C_{F_{\alpha_0}}$ is the nominal cornering stiffness, and the shape, peak, and curvature factors are also substituted as $C = C_y$, $D = F_{z_0}$ and $E = E_y$, respectively, and where C_y and E_y are the specific values of the shape and curvature factors for lateral force calculation.

Similarly, the nominal self-aligning torque $M_{S_0}^-(\alpha)$ results from Eq. (2.39) by the following substitutions. As the pneumatic trail establishes the relationship between the lateral force and the self-aligning torque (see Fig. 2.4), the stiffness factor for the self-aligning torque follows from Eq. (2.40) multiplying it by $(-t_p)$, i.e.,

$$B = -t_p \frac{C_{F_{\alpha_0}}}{CD}, \qquad (2.41)$$

where $t_p \approx c_4 l_c/2$ is an empirical approximation of the pneumatic trail with the empirical parameter c_4. The corresponding substitutions in Eq. (2.41) follow as $C = C_z$, $D = c_3 l_c F_{z_0}/2$ and $E = E_z$, where C_z and E_z are the self-aligning torque shape and curvature factors, respectively, and c_3 is an empirical parameter. A generalised description of these substitutions may also be found in PACEJKA (2006, p. 161).

Tyre characteristics depend on the vertical wheel load F_z (PACEJKA, 2006, pp. 4–6). It has been observed, however, that the tyre characteristics always have a typical shape, regardless of the actual wheel load. From this follows, that the nominal tyre characteristics $F_{y_0}^-$ and $M_{S_0}^-$ may be scaled to the actual load F_z by similarity approximations (PACEJKA, 2006, p. 157). This can be done in three steps:

- the lateral force may be scaled by the ratio F_z/F_{z_0} as

$$F_y^-(\alpha, F_z) = \frac{F_z}{F_{z_0}} F_{y_0}^-(\alpha); \qquad (2.42)$$

- as the previous step distorts the initial gradient, i.e., $\mathrm{d}F_y^-/\mathrm{d}\alpha|_{\alpha=0} \neq C_{F_{\alpha_0}}$, it has to be corrected by substituting the equivalent sideslip angle $\alpha_{eq} = \alpha F_{z_0}/F_z$ into Eq. (2.42) as $F_y^-(\alpha_{eq}, F_z)$ keeping the cornering stiffness unchanged;
- finally, the cornering stiffness may be set to a desired value by further scaling the sideslip angle with the ratio $C_{F_\alpha}(F_z)/C_{F_{\alpha_0}}$ as

$$\alpha_{eq}(\alpha, F_z) = \frac{F_{z_0}}{F_z} \frac{C_{F_\alpha}(F_z)}{C_{F_{\alpha_0}}} \alpha, \qquad (2.43)$$

where

$$C_{F_\alpha}(F_z) = c_1 c_2 F_{z_0} \sin\left(2 \arctan\left(\frac{F_z}{c_2 F_{z_0}}\right)\right) \qquad (2.44)$$

is the empirical expression of the cornering stiffness as a function of F_z, c_1 and c_2 are empirical parameters, and $C_{F_{\alpha_0}} \equiv C_{F_\alpha}(F_{z_0})$.

The self-aligning torque may be scaled according to a similar logic and by substituting Eq. (2.43) into $M_{S_0}^-$ as

$$M_S^-(\alpha, F_z) = \frac{F_z}{F_{z_0}} \frac{C_{M_\alpha}(F_z)}{C_{M_{\alpha_0}}} \frac{C_{F_{\alpha_0}}}{C_{F_\alpha}(F_z)} M_{S_0}^-(\alpha_{eq}(\alpha, F_z)), \qquad (2.45)$$

where

$$C_{M_\alpha}(F_z) = c_4 \frac{l_c}{2} \sqrt{\frac{F_z}{F_{z_0}}} C_{F_\alpha}(F_z) \qquad (2.46)$$

is the empirical expression of the aligning stiffness as a function of F_z, and $C_{M_{\alpha_0}} \equiv C_{M_\alpha}(F_{z_0})$. For a detailed description, refer to PACEJKA (2006, pp. 158–160).

In addition to the vertical load, the longitudinal force also influences the tyre characteristics as the friction circle of the tyre limits the vector sum of tyre forces in the contact patch, i.e., $\sqrt{F_x^2 + F_y^2} \leq \mu F_z$. Tyre models handle this by combining longitudinal and lateral force characteristics into a unified, combined force characteristic. Here, the rolling motions are not modelled and thus the longitudinal slip and the longitudinal force characteristics are not available. However, we may take the longitudinal forces into account as explicitly given input variables (PACEJKA, 2006, pp. 170–172). This simplification particularly fits to the state-of-the-art in-wheel motors as they can change their torque from the negative to the positive peak within less than 10 ms. Thus, only a negligible delay is caused by the dynamics of the electric motors, and they are almost ideal input sources of the longitudinal tyre forces.

In order to simultaneously include the load and the longitudinal forces into the the tyre characteristics, we may introduce the empirical scaling factor

$$\varphi_n(F_x, F_z) = \left[1 - \left(\frac{F_x}{F_z} \right)^n \right]^{\frac{1}{n}}, \; n \in \mathbb{N}. \qquad (2.47)$$

The aim of the scaling factor is to adjust the lateral force or the self-aligning moment capacities by adjusting the radius of the friction circle or limiting the vector sum of tyre forces if the load or the longitudinal force changes, respectively. We may then apply the scaling factor (2.47) with various exponents n to Eqns. (2.44) and (2.46), and extend Eq. (2.44) with longitudinal force terms, resulting in

$$C_{F_\alpha}(F_x, F_z) = \varphi_4(F_x, F_z)\left(c_1 c_2 F_{z_0} \sin\left(2\arctan\left(\frac{F_z}{c_2 F_{z_0}}\right)\right) - \frac{F_z}{2}\right)$$
$$+ \frac{F_z - F_x}{2} \tag{2.48}$$

and

$$C_{M_\alpha}(F_x, F_z) = \varphi_2^2(F_x, F_z)\, c_4 \frac{l_c}{2}\sqrt{\frac{F_z}{F_{z_0}}} C_{F_\alpha}(F_z). \tag{2.49}$$

Substituting Eq. (2.48) into (2.43) and applying the scaling factor (2.47) results in a scaled sideslip angle

$$\alpha_{eq}(\alpha, F_x, F_z) = \frac{1}{\varphi_2(F_x, F_z)} \frac{C_{F_\alpha}(F_x, F_z) F_{z_0}}{C_{F_{\alpha_0}}} \alpha. \tag{2.50}$$

Eventually, we may also apply the scaling factor (2.47) to (2.42) and substitute (2.50) resulting in the scaled and adjusted lateral force

$$F_y^-(\alpha, F_x, F_z) = \varphi_2(F_x, F_z)\frac{F_z}{F_{z_0}} F_{y_0}^-(\alpha_{eq}(\alpha, F_x, F_z)). \tag{2.51}$$

Similarly, we may also scale Eq. (2.45) resulting in the scaled and adjusted self-aligning torque

$$M_S^-(\alpha, F_x, F_z) =$$
$$\varphi_2(F_x, F_z)\frac{F_z}{F_{z_0}}\frac{C_{M_\alpha}(F_x, F_z)}{C_{M_{\alpha_0}}}\frac{C_{F_{\alpha_0}}}{C_{F_\alpha}(F_x, F_z)} M_{S_0}^-(\alpha_{eq}(\alpha, F_x, F_z))$$
$$- \frac{c_9 l_c F_x F_y^-(\alpha, F_x, F_z)}{2F_{z_0}} - \frac{c_{10} l_c F_x}{2}, \tag{2.52}$$

where the second and third terms are added to capture the self-aligning torque produced by the longitudinal force as a result of the contact patch distortion of the lateral force and an initial offset, respectively, and where c_9 and c_{10} are empirical parameters. By substituting the corresponding quantities, Eqns. (2.48)–(2.52) may be calculated for each wheel as functions of actual sideslip angle α_i, tyre load F_{z_i} and traction force F_{x_i} from in-wheel motors, i.e., $F_{y_i} = -F_y^-(\alpha_i, F_{x_i}, F_{z_i})$, $M_{S_i} = -M_S^-(\alpha_i, F_{x_i}, F_{z_i})$, $C_{F_{\alpha_i}} = C_{F_\alpha}(F_{x_i}, F_{z_i})$ and $C_{M_{\alpha_i}} = C_{M_\alpha}(F_{x_i}, F_{z_i})$ where $i = 1\ldots4$. The actual tyre characteristics are shown in Fig. 2.5 while the tyre parameters are given in Table 5.1d.

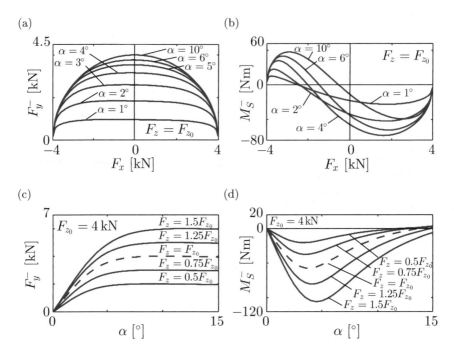

Figure 2.5: Combined characteristics of longitudinal tyre force F_x with (a) lateral tyre force F_y^- and (b) self-aligning torque M_S^-; further, load influence on (c) lateral tyre force F_y^- and (d) self-aligning torque M_S^-

2.3.3 Bore Torque Modelling

Turning the wheel about an axis normal to the contact patch forces some tread particles of the tyre to start sliding. This rotary motion is called bore motion by HIRSCHBERG ET AL. (2007), or alternatively turn by PACEJKA (2006, pp. 4, 64–69). The slide generates a counteracting torque about the normal axis called bore torque. The bore torque model of HIRSCHBERG ET AL. (2007) is based on a simple approximation of sliding motions and approximates the contact patch with a circle where the motions of the tread elements can be calculated easily from Fig. 2.6a. The radius $r_P = (l_c + w_c)/4$ of the approximating circle results as mean value from length l_c and width w_c of the contact patch. The turning speed ω_n and the wheel center velocity approximated by the vehicle velocity v

determine the slip

$$s_B(r, \omega_n) = \frac{r\omega_n}{r_D \, \|\mathbf{\Omega}\|} \approx \frac{r\omega_n}{v} \qquad (2.53)$$

of an infinitesimal patch element described by r and ϕ, see Fig. 2.6a, where the rolling velocity of the wheel is approximately $\|\mathbf{\Omega}\| \approx v/r_D$ and r_D is the dynamic rolling radius of the wheel. For small slip, the tangential force F_B may be approximated by multiplying s_B with a stiffness coefficient resulting from the tyre characteristics (HIRSCHBERG ET AL., 2007). Here, the cornering stiffness (2.44) is used generating the circumreferential force

$$F_B(r, \omega_n, F_z) = s_B(r, \omega_n) \, C_{F_\alpha}(F_z), \qquad (2.54)$$

where the influence of the longitudinal force is neglected for the sake of simplicity. As the torque generated by an individual patch element results as rF_B, the overall bore torque may be obtained by integration over the circularly approximated area of the contact patch:

$$
\begin{aligned}
M_B^*(\omega_n, F_z) &= -\int_A rF_B(r, \omega_n, F_z)\frac{\mathrm{d}A}{A} = -\frac{1}{A}\int_A rF_B(r, \omega_n, F_z)\,\mathrm{d}A \\
&= -\frac{1}{A}\int_A rs_B(r, \omega_n)C_{F_\alpha}(F_z)\,\mathrm{d}A \\
&= -\frac{1}{A}\int_A r\frac{r\omega_n}{v}C_{F_\alpha}(F_z)\,\mathrm{d}A = -\frac{\omega_n C_{F_\alpha}(F_z)}{vA}\int_A r^2\,\mathrm{d}A \\
&= -\frac{\omega_n C_{F_\alpha}(F_z)}{v\pi r_P^2}\int_0^{r_P}\int_0^{2\pi} r^3\,\mathrm{d}\phi\,\mathrm{d}r = -\frac{\omega_n C_{F_\alpha}(F_z)r_P^2}{2v},
\end{aligned}
$$
$$(2.55)$$

where $\mathrm{d}A = r\,\mathrm{d}\phi\,\mathrm{d}r$ and the division by $A = \pi r_P^2$ renders Eq. (2.55) to result in torque. However, for high slip, all tread elements are sliding and the tangential force $F_B = \mu F_z$ results from the load F_z and friction coefficient μ, resulting in

$$M_B^{\max}(F_z) = \int_A r\mu F_z\frac{\mathrm{d}A}{A} = \frac{\mu F_z}{A}\int_A r\,\mathrm{d}A = \frac{\mu F_z}{\pi r_P^2}\int_0^{r_P}\int_0^{2\pi} r^2\,\mathrm{d}\phi\,\mathrm{d}r = \frac{2}{3}r_P\mu F_z.$$
$$(2.56)$$

This limits the actual bore torque to

$$M_B = -\operatorname{sgn}(\omega_n)\min\{|M_B^*|, M_B^{\max}\}, \qquad (2.57)$$

which is a piecewise function leading to difficulties during simulation by numerical integration of the EoM. Therefore, it is approximated by a smooth

(a)   (b)

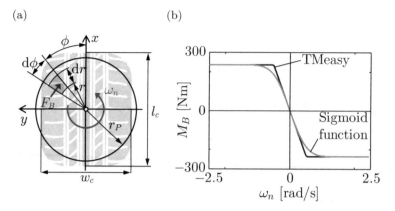

Figure 2.6: Bore torque generation: (a) motions and forces and (b) bore torque characteristics for $F_z = F_{z_0}$, $v = 1\,\mathrm{km/h}$ and parameters of Table 5.1

sigmoid function, the Gauss error function ($\mathrm{erf}(\cdot) = (2/\sqrt{\pi}) \int_0^{(\cdot)} e^{-x^2}\,\mathrm{d}x$, see ANDREWS, 1992, p. 110) as

$$M_B\left(\omega_n, F_z\right) = M_B^{\max}(F_z)\,\mathrm{erf}\left(c\omega_n\right), \tag{2.58}$$

where c is a scaling factor adjusting the gradient of Eq. (2.58) to that of Eq. (2.55), i.e.,

$$\left[\frac{\partial}{\partial \omega_n}\left(M_B^{\max}(F_z)\,\mathrm{erf}\left(c\omega_n\right)\right)\right]\bigg|_{\omega_n=0} = M_B^{\max}(F_z)c\frac{2}{\sqrt{\pi}}e^{-c^2\omega_n^2}\bigg|_{\omega_n=0}$$

$$= M_B^{\max}(F_z)c\frac{2}{\sqrt{\pi}} \overset{!}{=} \frac{\partial M_B^*(\omega_n, F_z)}{\partial \omega_n} = -\frac{C_{F_\alpha}(F_z)r_P^2}{2v}, \tag{2.59}$$

from which c follows as

$$c = -\frac{C_{F_\alpha}(F_z)r_P^2\sqrt{\pi}}{4vM_B^{\max}(F_z)}. \tag{2.60}$$

The difference between the bore torque characteristic (2.58) and the original TMeasy model can be seen in Fig. 2.6b. By substituting the corresponding quantities, the bore torque may be calculated for each wheel as function of the actual wheel load F_{z_i} and its normal turning velocity, i.e., $M_{B_i} = M_B\left(\omega_{n_i}, F_{z_i}\right)$ where $\omega_{n_i} = \omega + \dot{\delta}_i$.

2.3.4 Load Distribution and Load Transfer

The wheel loads F_{z_i} are generally unequal, firstly due to the unequal static mass distribution of the vehicle for $l_f \neq l_r$, and secondly due to the dynamic load transfer during cornering. In order to estimate the individual wheel loads without considering suspension dynamics, a simple steady-state model of the lateral load transfer may be used (PAUWELUSSEN, 2014, pp. 141–142).

By assuming symmetry of the vehicle about its longitudinal x_0-axis, the difference of wheel loads between left and right sides follows from the lateral load transfer only, resulting from the lateral acceleration. The lateral acceleration is obtained from projecting $\boldsymbol{a}_0 = \mathrm{d}\boldsymbol{v}_0/\mathrm{d}t$ into the body fixed frame K_0 as

$$\boldsymbol{a}_0' = \boldsymbol{S}_{I0}^{-1}\boldsymbol{a}_0 = \boldsymbol{S}_{I0}^{\mathrm{T}}\boldsymbol{a}_0 = \begin{bmatrix} -v(\dot{\beta}+\omega)\sin\beta & v(\dot{\beta}+\omega)\cos\beta & 0 \end{bmatrix}^{\mathrm{T}}, \quad (2.61)$$

see also Eqns. (2.4) and (A.7). For the sake of simplicity, only steady-state cornering is considered here by applying $\beta = \mathrm{const.}$, $\omega = \mathrm{const.}$ Thus, the y-component of Eq. (2.61) results in the steady-state lateral acceleration

$$a_{0_{y\infty}}' = v\omega\cos\beta. \quad (2.62)$$

Assuming that wheel turning angles δ_i are small resulting in $\cos\delta_i \approx 1$, the sum of lateral forces acting during cornering may be approximated as

$$\sum_{i=1\ldots4} F_{y_i} \approx m a_{0_y}' \quad (2.63)$$

compensating the inertia force $(-ma_{0_y}')$ resulting from the lateral acceleration a_{0_y}' where $m = m_b + 4m_w$ is the total vehicle mass. Further, $\sum_{i=1\ldots4} F_{y_i}$ has to fulfil the equilibrium of moments about the x_0-axis with respect to the CoG as

$$h \sum_{i=1\ldots4} F_{y_i} + (F_{z_1} + F_{z_3} - F_{z_2} - F_{z_4})\frac{l_t}{2} \approx 0, \quad (2.64)$$

where h is the height of the vehicle's CoG and $l_t = 2a + r_f + r_r$ is the average track width, see Fig. 2.7a. Averaging the track width is for simplification. In case of cornering there is a load difference

$$2\Delta F_z = F_{z_2} + F_{z_4} - F_{z_1} - F_{z_3} \quad (2.65)$$

between the left and right sides of the vehicle. Substitution of Eqns. (2.63) and (2.65) into Eq. (2.64) yields

$$\Delta F_z = \frac{hma_{0_y}'}{l_t}. \quad (2.66)$$

In the longitudinal direction, only the static load distribution is considered following from Fig. 2.7b and resulting in the force and torque equilibriums

$$F_{z_f} + F_{z_r} - mg = 0, \quad F_{z_r}(l_f + l_r) - mgl_f = 0. \tag{2.67}$$

This yields front and rear axle loads as

$$F_{z_f} = \frac{mgl_r}{l_f + l_r} \quad \text{and} \quad F_{z_r} = \frac{mgl_f}{l_f + l_r}. \tag{2.68}$$

By using the same front–rear distribution ratios $l_r/(l_f+l_r)$ and $l_f/(l_f+l_r)$ for the lateral load transfer distribution, the vertical load for each wheel may be roughly estimated from the combination of Eqns. (2.66) and (2.68) as

$$
\begin{aligned}
F_{z_1,z_2} &= \frac{F_{z_f}}{2} \mp \Delta F_z \frac{l_r}{l_f + l_r} = m\left(\frac{g}{2} \mp \frac{ha'_{0y\infty}}{l_t}\right)\frac{l_r}{l_f + l_r}, \\
F_{z_3,z_4} &= \frac{F_{z_r}}{2} \mp \Delta F_z \frac{l_f}{l_f + l_r} = m\left(\frac{g}{2} \mp \frac{ha'_{0y\infty}}{l_t}\right)\frac{l_f}{l_f + l_r}.
\end{aligned}
\tag{2.69}
$$

(a)   (b)

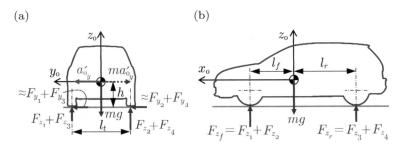

Figure 2.7: Lateral acceleration and forces during cornering (a) and static load distribution (b)

3 Symbolic Linearisation of Equations of Motion

Usually, equations of motion (EoM) of multi-body systems are highly nonlinear. This, however, may be disadvantageous in certain applications like modal analysis, stability analysis, sensitivity analysis, computation of input-output transfer functions, or generally to gain better understanding of the system as proposed by GONZÁLEZ ET AL. (2017). According to BAPST ET AL. (2014) and BUSCH (2015), linearised models have lower computational complexity than their nonlinear counterparts and, therefore, may be used for speeding up optimisation. One of the most common reasons for linearisation is control design where linear control methods are popular because they are well understood, proven and easy-to-use.

There exist many studies dealing with linearisation, however, predominantly based on a numerical approach. The nature of these methods strongly depends on how the EoM is formulated (e.g., in terms of Cartesian or generalised coordinates) and type of constraints (closed-loop kinematics and/or non-holonomic constraints). Sometimes, the better approach is to obtain the linearised system equations in symbolic form. Symbolic representation has the obvious advantage that the often cumbersome procedure of linearisation has to be performed only once and any forthcoming parameter variations may be done directly based on the linearised equations. This property may be especially advantageous in case of optimisation or control design problems (e.g., gain scheduling).

An early approach for symbolic linearisation of multi-body systems was reported by NEUMAN AND MURRAY (1984) for unconstrained systems. GE ET AL. (2005) proposed a symbolic linearisation method by using Cartesian coordinates, where non-holonomic constraints are not discussed. TULPULE (2014) presented a method requiring coordinate partitioning of Cartesian coordinates, where some computational issues are discussed as well. PETERSON ET AL. (2015) introduced the most elaborated method for systems formalised by Kane's equations, which allows the use of any generalised coordinates. The system may contain kinematic loops as well as linear non-holonomic constraints. A common property of the methods from NEGRUT AND ORTIZ (2006), TULPULE (2014) and PETERSON ET AL. (2015) is that they require a coordinate partitioning prior to linearisation and none deal with nonlinear non-holonomic constraints.

In this chapter, a computationally efficient, control-oriented symbolic linearisation method is proposed for non-holonomic multi-body systems with closed-loop kinematics. The system may be described by any generalised coordinates and velocities. Important properties of the method are that no early partitioning of the coordinates is required, and the non-holonomic constraints may be either linear or nonlinear. The chapter is organised as follows. The symbolic Taylor expansion of the EoM is discussed first as the core of the linearisation procedure. Subsequently, a state reduction method is introduced to eliminate the redundant coordinates, and the frequency domain representation of the linearised system is discussed as well. Finally, the application of the linearisation method to the vehicle model is presented.

3.1 Symbolic Taylor Expansion

The goal is to obtain the linearised EoM in the classical form as ordinary differential equations (ODE) instead of DAE. Therefore, we first need to eliminate the Lagrange multipliers $\boldsymbol{\lambda}$ in order to get a set of ordinary differential equations, and then we may perform the Taylor expansion.

Following LAULUSA AND BAUCHAU (2008), let us rewrite Eq. (2.26) as

$$\dot{\boldsymbol{z}} = \boldsymbol{M}^{-1}\hat{\boldsymbol{q}} + \boldsymbol{M}^{-1}\boldsymbol{C}_{\boldsymbol{z}}^{\mathrm{T}}\boldsymbol{\lambda}, \tag{3.1}$$

where $\hat{\boldsymbol{q}} = \boldsymbol{q} - \boldsymbol{k}$. Substitution into Eq. (2.34) and rearrangement yields

$$\boldsymbol{\lambda} = -\left(\boldsymbol{C}_{\boldsymbol{z}}\boldsymbol{M}^{-1}\boldsymbol{C}_{\boldsymbol{z}}^{\mathrm{T}}\right)^{-1}\left(\boldsymbol{C}_{\boldsymbol{z}}\boldsymbol{M}^{-1}\hat{\boldsymbol{q}} + \boldsymbol{\gamma}\right). \tag{3.2}$$

Back-substitution into Eq. (3.1) finally results in the ODE

$$\dot{\boldsymbol{z}} = \boldsymbol{\zeta} := \boldsymbol{M}^{-1}\hat{\boldsymbol{q}} - \boldsymbol{M}^{-1}\boldsymbol{C}_{\boldsymbol{z}}^{\mathrm{T}}\left(\boldsymbol{C}_{\boldsymbol{z}}\boldsymbol{M}^{-1}\boldsymbol{C}_{\boldsymbol{z}}^{\mathrm{T}}\right)^{-1}\left(\boldsymbol{C}_{\boldsymbol{z}}\boldsymbol{M}^{-1}\hat{\boldsymbol{q}} + \boldsymbol{\gamma}\right). \tag{3.3}$$

Let us define $\tilde{\boldsymbol{x}} = \begin{bmatrix} \boldsymbol{y}^{\mathrm{T}} & \boldsymbol{z}^{\mathrm{T}} \end{bmatrix}^{\mathrm{T}}$ as pseudo-state vector and $\boldsymbol{u}$ as input vector, which typically consists of controllable forces being part of vector $\boldsymbol{q}$ in

Eq. (2.29) and thus of vector $\hat{\boldsymbol{q}}$ in Eq. (3.3). In our case, $\boldsymbol{u}$ summarises the traction forces F_{x_i}, $i = 1 \ldots 4$, in Fig. 2.1. Note that $\tilde{\boldsymbol{x}}$ may contain redundant variables due to loop closing constraints, which is why it cannot be regarded as a classical state vector.

Using these definitions, Eqns. (2.3) and (3.3) may be arranged as pseudo-state space representation

$$\dot{\tilde{\boldsymbol{x}}} = \begin{bmatrix} \dot{\boldsymbol{y}} \\ \dot{\boldsymbol{z}} \end{bmatrix} = \begin{bmatrix} \boldsymbol{f}_v\,(\tilde{\boldsymbol{x}}, t) \\ \boldsymbol{\zeta}\,(\tilde{\boldsymbol{x}}, \boldsymbol{u}, t) \end{bmatrix} \tag{3.4}$$

and linearised by Taylor expansion as

$$\begin{aligned} &\frac{\mathrm{d}}{\mathrm{d}t}\,(\tilde{\boldsymbol{x}}_0 + \delta\tilde{\boldsymbol{x}}) \\ &\approx \begin{bmatrix} \boldsymbol{f}_v\,(\tilde{\boldsymbol{x}}_0) \\ \boldsymbol{\zeta}\,(\tilde{\boldsymbol{x}}_0, \boldsymbol{u}_0) \end{bmatrix} + \frac{\partial}{\partial\tilde{\boldsymbol{x}}}\begin{bmatrix} \boldsymbol{f}_v \\ \boldsymbol{\zeta} \end{bmatrix}\bigg|_{\substack{\tilde{\boldsymbol{x}}=\tilde{\boldsymbol{x}}_o \\ \boldsymbol{u}=\boldsymbol{u}_o}}\delta\tilde{\boldsymbol{x}} + \frac{\partial}{\partial\boldsymbol{u}}\begin{bmatrix} \boldsymbol{f}_v \\ \boldsymbol{\zeta} \end{bmatrix}\bigg|_{\substack{\tilde{\boldsymbol{x}}=\tilde{\boldsymbol{x}}_o \\ \boldsymbol{u}=\boldsymbol{u}_o}}\delta\boldsymbol{u}, \end{aligned} \tag{3.5}$$

where $\tilde{\boldsymbol{x}}_o$ and $\boldsymbol{u}_o$ are state variables and input values at the desired operating point or trajectory, and $\delta\tilde{\boldsymbol{x}} = \tilde{\boldsymbol{x}} - \tilde{\boldsymbol{x}}_0$ and $\delta\boldsymbol{u} = \boldsymbol{u} - \boldsymbol{u}_0$ are small deviations of $\tilde{\boldsymbol{x}}$ and $\boldsymbol{u}$ from $\tilde{\boldsymbol{x}}_0, \boldsymbol{u}_0$, respectively. Assuming that the operating point fulfills Eq. (3.4), i.e., $\dot{\tilde{\boldsymbol{x}}}_0 = \begin{bmatrix} \boldsymbol{f}_v\,(\tilde{\boldsymbol{x}}_0) & \boldsymbol{\zeta}\,(\tilde{\boldsymbol{x}}_0, \boldsymbol{u}_0) \end{bmatrix}^{\mathrm{T}}$, we may deduce the linearised EoM

$$\delta\dot{\tilde{\boldsymbol{x}}} \approx \begin{bmatrix} \boldsymbol{A}_{\mathrm{I}} \\ \boldsymbol{A}_{\mathrm{II}} \end{bmatrix}\delta\tilde{\boldsymbol{x}} + \begin{bmatrix} \boldsymbol{B}_{\mathrm{I}} \\ \boldsymbol{B}_{\mathrm{II}} \end{bmatrix}\delta\boldsymbol{u} \tag{3.6}$$

from Eq. (3.5), where the coefficient matrices may be expressed as

$$\boldsymbol{A}_{\mathrm{I}} = \frac{\partial\boldsymbol{f}_v}{\partial\tilde{\boldsymbol{x}}}\bigg|_{\substack{\tilde{\boldsymbol{x}}=\tilde{\boldsymbol{x}}_o \\ \boldsymbol{u}=\boldsymbol{u}_o}}, \quad \boldsymbol{B}_{\mathrm{I}} = \frac{\partial\boldsymbol{f}_v}{\partial\boldsymbol{u}}\bigg|_{\substack{\tilde{\boldsymbol{x}}=\tilde{\boldsymbol{x}}_o \\ \boldsymbol{u}=\boldsymbol{u}_o}}, \tag{3.7}$$

$$\boldsymbol{A}_{\mathrm{II}} = \frac{\partial\boldsymbol{\zeta}}{\partial\tilde{\boldsymbol{x}}}\bigg|_{\substack{\tilde{\boldsymbol{x}}=\tilde{\boldsymbol{x}}_o \\ \boldsymbol{u}=\boldsymbol{u}_o}}, \quad \boldsymbol{B}_{\mathrm{II}} = \frac{\partial\boldsymbol{\zeta}}{\partial\boldsymbol{u}}\bigg|_{\substack{\tilde{\boldsymbol{x}}=\tilde{\boldsymbol{x}}_o \\ \boldsymbol{u}=\boldsymbol{u}_o}}. \tag{3.8}$$

Due to $\boldsymbol{f}_v \neq \boldsymbol{f}_v(\boldsymbol{u})$, Eqns. (2.3) and (3.7) yield $\boldsymbol{B}_{\mathrm{I}} = \boldsymbol{0}$. For simplicity, let us apply the abbreviation $(\cdot)\big|_{\substack{\tilde{\boldsymbol{x}}=\tilde{\boldsymbol{x}}_o \\ \boldsymbol{u}=\boldsymbol{u}_o}} =: (\cdot)_*$ in the following.

Practical implementation of Eqns. (3.1)–(3.8) may imply some computational issues, especially the inversion of $\boldsymbol{M}$ and $\boldsymbol{C}_z\boldsymbol{M}^{-1}\boldsymbol{C}_z^{\mathrm{T}}$ in Eq. (3.3). Usually, these are large matrices with complicated expressions in symbolic form, where calculation of inverses with computer algebra software may cause memory overflow even in relatively simple cases like the vehicle model in the present work, see also the detailed description of $\boldsymbol{M}$ and $\boldsymbol{C_z}$ in the Appendix. Therefore, coefficient matrices $\boldsymbol{A}_{\text{II}}$ and $\boldsymbol{B}_{\text{II}}$ cannot be calculated directly according to Eq. (3.8). Instead, the following procedure may be applied.

Let us define

$$\widetilde{\boldsymbol{M}} := \boldsymbol{C_z} \boldsymbol{M}^{-1} \boldsymbol{C_z^{\text{T}}}, \tag{3.9}$$

$$\tilde{\gamma} := \boldsymbol{C_z} \boldsymbol{M}^{-1} \hat{\boldsymbol{q}} + \gamma \tag{3.10}$$

reducing Eq. (3.3) to

$$\zeta := \boldsymbol{M}^{-1} \hat{\boldsymbol{q}} - \boldsymbol{M}^{-1} \boldsymbol{C_z^{\text{T}}} \widetilde{\boldsymbol{M}}^{-1} \tilde{\gamma}. \tag{3.11}$$

Then, $\boldsymbol{A}_{\text{II}}$ in Eq. (3.8) may be expressed as

$$\boldsymbol{A}_{\text{II}} = \frac{\partial \left(\boldsymbol{M}^{-1} \hat{\boldsymbol{q}} \right)}{\partial \tilde{\boldsymbol{x}}} \bigg|_* - \frac{\partial \left(\boldsymbol{M}^{-1} \boldsymbol{C_z^{\text{T}}} \widetilde{\boldsymbol{M}}^{-1} \tilde{\gamma} \right)}{\partial \tilde{\boldsymbol{x}}} \bigg|_* . \tag{3.12}$$

It is necessary to reformulate (3.12) such that direct symbolic calculation of $\boldsymbol{M}^{-1}$ and its derivatives is avoided. In the following, let the differentiation be performed element-wise with respect to $\tilde{x}_i$. From the identity $\boldsymbol{M}^{-1} \boldsymbol{M} = \boldsymbol{I}$, we get

$$\frac{\partial \left(\boldsymbol{M}^{-1} \boldsymbol{M} \right)}{\partial \tilde{x}_i} = \frac{\partial \boldsymbol{I}}{\partial \tilde{x}_i} \equiv \boldsymbol{0} \tag{3.13}$$

or

$$\frac{\partial \boldsymbol{M}^{-1}}{\partial \tilde{x}_i} \boldsymbol{M} + \boldsymbol{M}^{-1} \frac{\partial \boldsymbol{M}}{\partial \tilde{x}_i} = \boldsymbol{0}. \tag{3.14}$$

After rearrangement we obtain

$$\frac{\partial \boldsymbol{M}^{-1}}{\partial \tilde{x}_i} = -\boldsymbol{M}^{-1} \frac{\partial \boldsymbol{M}}{\partial \tilde{x}_i} \boldsymbol{M}^{-1}. \tag{3.15}$$

Considering the identity $(\boldsymbol{M}^{-1})_* = (\boldsymbol{M}_*)^{-1}$, we get

$$\boldsymbol{M}_* := \frac{\partial \boldsymbol{M}^{-1}}{\partial \tilde{x}_i} \bigg|_* = -\boldsymbol{M}_*^{-1} \frac{\partial \boldsymbol{M}}{\partial \tilde{x}_i} \bigg|_* \boldsymbol{M}_*^{-1}. \tag{3.16}$$

By this, the direct symbolic calculation of $\boldsymbol{M}^{-1}$ and subsequent differentiation is avoided. The computational burden of finding $\boldsymbol{M}_*^{-1}$ for given $\tilde{\boldsymbol{x}}_0$ is significantly lower than that of $\boldsymbol{M}^{-1}$ as general function of $\tilde{\boldsymbol{x}}$, since the substitution of the operating point simplifies many expressions in $\boldsymbol{M}$ to numbers or even pure numerical inversion may be sufficient.

By utilising Eq. (3.16) and the product rule consequently, the first term of Eq. (3.12) follows as

$$
\begin{aligned}
\frac{\partial \left(M^{-1}\hat{q}\right)}{\partial \tilde{x}_i}\bigg|_* &= \left(\frac{\partial M^{-1}}{\partial \tilde{x}_i}\hat{q} + M^{-1}\frac{\partial \hat{q}}{\partial \tilde{x}_i}\right)\bigg|_* \\
&= \mathcal{M}_*\hat{q}_* + M_*^{-1}\frac{\partial \hat{q}}{\partial \tilde{x}_i}\bigg|_* .
\end{aligned}
\tag{3.17}
$$

For the second term in Eq. (3.12), let us firstly calculate intermediate derivatives of non-trivial terms. The rule (3.15), (3.16) also applies to derivatives of $\widetilde{M}^{-1}$, i.e.,

$$
\frac{\partial \widetilde{M}^{-1}}{\partial \tilde{x}_i}\bigg|_* = -\widetilde{M}_*^{-1}\frac{\partial \widetilde{M}}{\partial \tilde{x}_i}\bigg|_*\widetilde{M}_*^{-1} =: \widetilde{\mathcal{M}}_*,
\tag{3.18}
$$

where $\widetilde{M}_*^{-1} = \left(C_{z*}M_*^{-1}C_{z*}^\mathrm{T}\right)^{-1}$ according to Eq. (3.9) and

$$
\frac{\partial \widetilde{M}}{\partial \tilde{x}_i}\bigg|_* = \frac{\partial C_z}{\partial \tilde{x}_i}\bigg|_*M_*^{-1}C_{z*}^\mathrm{T} + C_{z*}\mathcal{M}_*C_{z*}^\mathrm{T} + C_{z*}M_*^{-1}\frac{\partial C_z^\mathrm{T}}{\partial \tilde{x}_i}\bigg|_*
\tag{3.19}
$$

by using substitution (3.16). The differentiation of (3.10) results in

$$
\frac{\partial \tilde{\gamma}}{\partial \tilde{x}_i}\bigg|_* = \frac{\partial C_z}{\partial \tilde{x}_i}\bigg|_*M_*^{-1}\hat{q}_* + C_{z*}\mathcal{M}_*\hat{q}_* + C_{z*}M_*^{-1}\frac{\partial \hat{q}}{\partial \tilde{x}_i}\bigg|_* + \frac{\partial \gamma}{\partial \tilde{x}_i}\bigg|_* .
\tag{3.20}
$$

By summarising Eqns. (3.17)–(3.20), the i-th column of matrix (3.12) reads as

$$
\begin{aligned}
a_{\mathrm{II}_i} &= \frac{\partial \left(M^{-1}\hat{q}\right)}{\partial \tilde{x}_i}\bigg|_* - \left(\frac{\partial M^{-1}}{\partial \tilde{x}_i}\bigg|_*C_{z*}^\mathrm{T}\widetilde{M}_*^{-1}\tilde{\gamma}_* + M_*^{-1}\frac{\partial C_z^\mathrm{T}}{\partial \tilde{x}_i}\bigg|_*\widetilde{M}_*^{-1}\tilde{\gamma}_* \right.\\
&\quad \left. + M_*^{-1}C_{z*}^\mathrm{T}\frac{\partial \widetilde{M}^{-1}}{\partial \tilde{x}_i}\bigg|_*\tilde{\gamma}_* + M_*^{-1}C_{z*}^\mathrm{T}\widetilde{M}_*^{-1}\frac{\partial \tilde{\gamma}}{\partial \tilde{x}_i}\bigg|_*\right)\\
&= \mathcal{M}_*\hat{q}_* + M_*^{-1}\frac{\partial \hat{q}}{\partial \tilde{x}_i}\bigg|_* - \mathcal{M}_*C_{z*}^\mathrm{T}\widetilde{M}_*^{-1}\tilde{\gamma}_* - M_*^{-1}\frac{\partial C_z^\mathrm{T}}{\partial \tilde{x}_i}\bigg|_*\widetilde{M}_*^{-1}\tilde{\gamma}_*\\
&\quad - M_*^{-1}C_{z*}^\mathrm{T}\widetilde{\mathcal{M}}_*\tilde{\gamma}_* - M_*^{-1}C_{z*}^\mathrm{T}\widetilde{M}_*^{-1}\frac{\partial \tilde{\gamma}}{\partial \tilde{x}_i}\bigg|_* .
\end{aligned}
\tag{3.21}
$$

The coefficient matrix may then be assembled as $A_{\mathrm{II}} = \begin{bmatrix} a_{\mathrm{II}_1} & \cdots & a_{\mathrm{II}_n} \end{bmatrix}$ where n is the number of elements of pseudo-state vector $\tilde{x}$.

The same procedure may be applied to the calculation of B_{II} in Eq. (3.8) except that the derivatives of ζ are calculated with respect to u. However, in our case inputs u have only influence on q resulting in $\hat{q} = \hat{q}(u)$. Therefore, derivation of Eq. (3.3) w.r.t. u in Eq. (3.8) reduces to

$$
\begin{aligned}
B_{II} &= \left[M^{-1}\frac{\partial\hat{q}}{\partial u} - M^{-1}C_z^{T}\left(C_z M^{-1}C_z^{T}\right)^{-1}C_z M^{-1}\frac{\partial\hat{q}}{\partial u} \right]\Bigg|_* \\
&= M_*^{-1}\left(I - C_{z*}^{T}\tilde{M}_*^{-1}C_{z*}M_*^{-1} \right)\frac{\partial\hat{q}}{\partial u}\Bigg|_* .
\end{aligned}
\tag{3.22}
$$

3.2 State Reduction

Equation (3.6) is only a pseudo-state-space representation, because the state variables $\tilde{x} \in \mathbb{R}^{f+g}$ are redundant, i.e., coupled by $2n_c$ constraints (2.10) and (2.22). In order to obtain a correct state-space representation of the system, it is necessary to eliminate the redundant coordinates and describe the system by independent coordinates x. This may be performed after the linearisation, simplifying the reduction procedure.

Due to the linearity of the system, the reduction may be performed by a linear state transformation expressed by the projection

$$
\delta x = T\,\delta\tilde{x},
\tag{3.23}
$$

where $\delta x \in \mathbb{R}^{f+g-2n_c}$ is a real state vector consisting of independent coordinates and velocities, and $T \in \mathbb{R}^{(f+g-2n_c)\times(f+g)}$ is an appropriate time-invariant projection matrix. The linearised position and velocity constraint equations result from Eqns. (2.10) and (2.22) as

$$
c(\tilde{x}_0 + \delta\tilde{x}) \approx c(\tilde{x}_0) + \frac{\partial c}{\partial\tilde{x}}\Bigg|_* \delta\tilde{x} \equiv \frac{\partial c}{\partial\tilde{x}}\Bigg|_* \delta\tilde{x} = 0
\tag{3.24}
$$

and

$$
\dot{c}(\tilde{x}_0 + \delta\tilde{x}) \approx \dot{c}(\tilde{x}_0) + \frac{\partial\dot{c}}{\partial\tilde{x}}\Bigg|_* \delta\tilde{x} \equiv \frac{\partial\dot{c}}{\partial\tilde{x}}\Bigg|_* \delta\tilde{x} = 0
\tag{3.25}
$$

by assuming $c(\tilde{x}_0) = 0$ and $\dot{c}(\tilde{x}_0) = 0$ for the reference point or trajectory. With the notations $C_{\tilde{x}} = \partial c/\partial\tilde{x}$ and $\dot{C}_{\tilde{x}} = \partial\dot{c}/\partial\tilde{x}$, we may assemble Eqns. (3.23)–(3.25) as linear system of equations

$$
\tilde{T}\,\delta\tilde{x} := \begin{bmatrix} T \\ C_{\tilde{x}}\big|_* \\ \dot{C}_{\tilde{x}}\big|_* \end{bmatrix} \delta\tilde{x} = \begin{bmatrix} \delta x \\ 0 \\ 0 \end{bmatrix} .
\tag{3.26}
$$

If T is chosen properly, the matrix $\tilde{T} \in \mathbb{R}^{(f+g)\times(f+g)}$ has full rank and may be inverted. Thus, the linearised pseudo-state $\delta\tilde{x}$ may be uniquely expressed by the state δx as

$$\delta\tilde{x} = \tilde{T}^{-1} \begin{bmatrix} \delta x \\ 0 \\ 0 \end{bmatrix} =: \overset{\approx}{T}\,\delta x, \qquad (3.27)$$

where $\overset{\approx}{T} \in \mathbb{R}^{(f+g)\times(f+g-2n_c)}$ summarises the first $f + g - 2n_c$ columns of $\tilde{T}^{-1}$ to be multiplied with δx. Multiplication of the linearised Eq. (3.6) by T yields

$$T\,\delta\dot{\tilde{x}} = T \begin{bmatrix} A_I \\ A_{II} \end{bmatrix} \delta\tilde{x} + T \begin{bmatrix} B_I \\ B_{II} \end{bmatrix} \delta u. \qquad (3.28)$$

By substituting the time derivative $\delta\dot{x} = T\delta\dot{\tilde{x}}$ of Eq. (3.23) for a time-invariant projection matrix T and Eq. (3.27), we may get the state-space description

$$\delta\dot{x} = T \begin{bmatrix} A_I \\ A_{II} \end{bmatrix} \overset{\approx}{T}\,\delta x + T \begin{bmatrix} B_I \\ B_{II} \end{bmatrix} \delta u$$

$$=: A\,\delta x + B\,\delta u, \qquad (3.29)$$

with system matrix $A \in \mathbb{R}^{n\times n}$ and input matrix $B \in \mathbb{R}^{n\times m}$, where $n = f + g - 2n_c = 7 + 6 - 2\cdot 2 = 9$ is the state space dimension and $m = 4$ is the number of inputs.

3.3 Representation in the Frequency Domain

Alternatively, linear systems may be represented in the frequency domain by using transfer functions. Transfer functions may result from a Laplace transformation of the quasi-time-invariant system (3.29) for $v \approx$ const., i.e.,

$$\mathscr{L}\{\delta\dot{x}\} \equiv s\mathscr{L}\{\delta x\} = A\mathscr{L}\{\delta x\} + B\mathscr{L}\{\delta u\}, \qquad (3.30)$$

where $s = \sigma + j\widehat{\omega}$ combines real part σ and imaginary part $j\widehat{\omega}$ with angular frequency $\widehat{\omega} = 2\pi f$, see OGATA (2010, pp. 33–35). By denoting Laplace transforms of δx and δu as $\delta x_{\mathscr{L}}(s) = \mathscr{L}\{\delta x(t)\}$ and $\delta u_{\mathscr{L}}(s) = \mathscr{L}\{\delta u(t)\}$ we get

$$\delta x_{\mathscr{L}} = (sI - A)^{-1} B\,\delta u_{\mathscr{L}} =: G_{\delta u \to \delta x}\,\delta u_{\mathscr{L}}, \qquad (3.31)$$

where $G_{\delta u \to \delta x}(s)$ is the transfer matrix characterising the connection between inputs δu and states δx.

3.4 Application to the Vehicle Model

In this section, the presented linearisation procedure is applied to the vehicle model. Nonlinearities in the vehicle model result from different sources. Firstly, the bodies perform rotational motions, which is why the description of their positions and velocities heavily involves trigonometric functions. Secondly, nonlinearities arise from the tyre forces and torques.

The point of interest for linearisation is trivial, i.e., $\tilde{x}_0 = 0$ and $u_0 = 0$. From this particular condition follows that $\delta\tilde{x} \equiv \tilde{x}$, $\delta u \equiv u$ and $\delta x \equiv x$. Thus, we may write the linear representations (3.29) and (3.31) in the simplified forms

$$\dot{x} = Ax + Bu \quad \text{and} \quad x_{\mathscr{L}} = G_{u \to x} u_{\mathscr{L}} \tag{3.32}$$

for the forthcoming calculations, where $x_{\mathscr{L}}$ and $u_{\mathscr{L}}$ are the Laplace transforms of x and u, respectively.

3.4.1 Symbolic Manipulations on the Vehicle Model

In the particular problem, position variables x_s, y_s and ψ are not relevant for control design and performance analysis, since the dynamics equations are decoupled from these variables. Thus, it is sufficient to omit these coordinates in position vector (2.1) and to reduce the pseudo-state vector in Eq. (3.4) here to

$$\tilde{x} = \begin{bmatrix} \delta_1 & \dots & \delta_4 & | & \beta & \omega & \dot{\delta}_1 & \dots & \dot{\delta}_4 \end{bmatrix}^{\mathrm{T}} \in \mathbb{R}^{10} \tag{3.33}$$

only. This definition truncates y and hence reduces the DoFs of the linear model to $f = 4$ while z and g remain unchanged, compare Eq. (3.33) to Eqns. (2.1), (2.2). The input vector is given by the longitudinal wheel forces $u = \begin{bmatrix} F_{x_1} & \dots & F_{x_4} \end{bmatrix}^{\mathrm{T}} \in \mathbb{R}^4$ resulting from wheel driving torques.

By using these definitions, Eq. (3.6) can be derived for the vehicle model. Although the procedure shown in Section 3.1 largely reduces the computational burden of linearisation, the resulting symbolic expressions are suitable only for use in computer algebra systems as the complete symbolic expressions are still so large that it is practically impossible to present them on paper down to the elementary details. However, some elements may be extracted to discuss and demonstrate the proposed linearisation approach.

For example, after deriving the EoM (2.36), we find for element $M^{(2,1)}$ of the mass matrix (2.27) by summing up the respective elements in (A.30)–(A.34)

$$
\begin{aligned}
M^{(2,1)} = m_w v \, [& t_f \left(\cos \left(\beta - \delta_1 \right) + \cos \left(\beta - \delta_2 \right) \right) \\
& + r_f \left(\sin \left(\beta - \delta_1 \right) - \sin \left(\beta - \delta_2 \right) \right) \\
& + t_r \left(\cos \left(\beta - \delta_3 \right) + \cos \left(\beta - \delta_4 \right) \right) \\
& + r_r \left(\sin \left(\beta - \delta_3 \right) - \sin \left(\beta - \delta_4 \right) \right) + 2(l_f - l_r) \cos \beta \,] .
\end{aligned}
\tag{3.34}
$$

Due to the computational difficulties mentioned in the previous section, it is hardly imaginable, that a 6×6 mass matrix with such elements can be inverted symbolically to be then differentiated with respect to $\tilde{x}_i$ as part of Eq. (3.12) to obtain the linearised coefficient matrix $\boldsymbol{A}_{\mathrm{II}}$.

However, by substituting $\tilde{\boldsymbol{x}} = \tilde{\boldsymbol{x}}_0 \equiv \boldsymbol{0}$ we may easily obtain for $\beta = \delta_i = 0$ that

$$
M_*^{(2,1)} = M^{(2,1)}|_{\tilde{\boldsymbol{x}} = \tilde{\boldsymbol{x}}_o} = 2 m_w v \left(t_f + t_r + l_f - l_r \right) .
\tag{3.35}
$$

Also, e.g., derivatives like

$$
\begin{aligned}
\left. \frac{\partial M^{(2,1)}}{\partial \delta_1} \right|_* &= \left[m_w v \left(t_f \sin \left(\beta - \delta_1 \right) - r_f \cos \left(\beta - \delta_1 \right) \right) \right]|_{\beta = \delta_i = 0} \\
&= -m_w v r_f
\end{aligned}
\tag{3.36}
$$

as part of Eq. (3.16) may be computed to finally come up with $\boldsymbol{\mathcal{M}}_*$ being used in (3.21) to compute the state matrix.

Besides the derivation of EoM (2.36) and the discussed computational issues in Eqns. (3.34)–(3.36), we need to account for the constraints and the state reduction, since the steering angles δ_i are coupled by the tie rods and are thus not independent. In order to eliminate the redundancy, we may define new state variables $\delta_f = (\delta_1 + \delta_2)/2$, $\delta_r = (\delta_3 + \delta_4)/2$, $\dot{\delta}_f = (\dot{\delta}_1 + \dot{\delta}_2)/2$ and $\dot{\delta}_r = (\dot{\delta}_3 + \dot{\delta}_4)/2$ as arithmetic means of left and right steering angles, respectively, and condense the pseudo-state vector (3.33) to

$$
\boldsymbol{x} = \begin{bmatrix} \delta_f & \delta_r & \beta & \omega & \dot{\delta}_f & \dot{\delta}_r \end{bmatrix}^{\mathrm{T}} \in \mathbb{R}^6 .
\tag{3.37}
$$

From these definitions, the 6×10-projection matrix in Eq. (3.23) results as

$$
\boldsymbol{T} =
\begin{bmatrix}
1/2 & 1/2 & 0 & 0 & \cdots & & & & & 0 \\
0 & 0 & 1/2 & 1/2 & 0 & \cdots & & & & 0 \\
0 & \cdots & & 0 & 1 & 0 & \cdots & & & 0 \\
0 & \cdots & & 0 & 1 & 0 & \cdots & & & 0 \\
0 & \cdots & & & 0 & 1/2 & 1/2 & 0 & 0 \\
0 & \cdots & & & & 0 & 0 & 1/2 & 1/2
\end{bmatrix},
\tag{3.38}
$$

which has to be supplemented by constraint information according to Eq. (3.26).

In order to obtain a better insight, let us investigate some details of the state reduction procedure starting from the constraint equation (2.10) of the front axle:

$$
c_1 = c_{1_x}^2 + c_{1_y}^2 - b^2 = 0,
\tag{3.39}
$$

where

$$
\begin{aligned}
c_{1_x} &= (h_f \cos \delta_2 + c \sin \delta_2) - (h_f \cos \delta_1 - c \sin \delta_1), \\
c_{1_y} &= 2a + (h_f \sin \delta_2 - c \cos \delta_2) - (h_f \sin \delta_1 + c \cos \delta_1), \\
c &= a - \frac{b}{2}
\end{aligned}
\tag{3.40}
$$

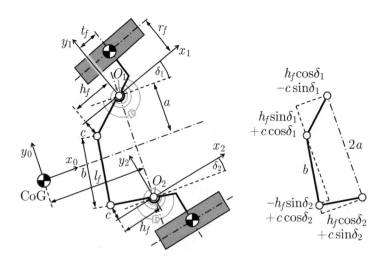

Figure 3.1: Front axle steering linkage

may either deduced from Eq. (2.10) or Fig. 3.1, see also Eq. (A.45). Only two partial derivatives of Eq. (3.39) with respect to $\tilde{\boldsymbol{x}}$ are not vanishing, i.e.,

$$\frac{\partial c_1}{\partial \delta_1}\bigg|_* = 2\left[c_{1_x}\left(h_f \sin \delta_1 + c \cos \delta_1\right) - c_{1_y}\left(h_f \cos \delta_1 - c \sin \delta_1\right)\right]\big|_{\delta_1=\delta_2=0}$$

$$= -2\left(2a - 2c\right)h_f \equiv -2bh_f,$$

$$(3.41)$$

$$\frac{\partial c_1}{\partial \delta_2}\bigg|_* = 2\left[c_{1_y}\left(h_f \cos \delta_2 + c \sin \delta_2\right) - c_{1_x}\left(h_f \sin \delta_2 - c \cos \delta_2\right)\right]\big|_{\delta_1=\delta_2=0}$$

$$= 2\left(2a - 2c\right)h_f \equiv 2bh_f,$$

$$(3.42)$$

where $c_{1_x}\big|_* = 0$ and $c_{1_y}\big|_* = 2a - 2c = b$ has been used. By using the corresponding variables and parameters $(\delta_3, \delta_4, h_r)$ for the rear axle, the second constraint equation $c_2 = 0$ in Eq. (2.10) and the non-vanishing terms may be obtained analogously as

$$\frac{\partial c_2}{\partial \delta_3}\bigg|_* = 2bh_r, \quad \frac{\partial c_2}{\partial \delta_4}\bigg|_* = -2bh_r \qquad (3.43)$$

finally resulting in the 2×10 constraint Jacobian

$$\boldsymbol{C}_{\tilde{\boldsymbol{x}}}\big|_* = \frac{\partial \boldsymbol{c}}{\partial \tilde{\boldsymbol{x}}}\bigg|_* = \begin{bmatrix} -2bh_f & 2bh_f & 0 & 0 & 0 & \cdots & 0 \\ 0 & 0 & 2bh_r & -2bh_r & 0 & \cdots & 0 \end{bmatrix} \qquad (3.44)$$

as part of $\widetilde{\boldsymbol{T}}$ in Eq. (3.26).

In order to compute the lower submatrix $\dot{\boldsymbol{C}}_{\tilde{\boldsymbol{x}}}\big|_*$ of $\widetilde{\boldsymbol{T}}$, we first have to differentiate constraint (3.39) with respect to time resulting in

$$\dot{c}_1 = 2c_{1_x}\dot{c}_{1_x} + 2c_{1_y}\dot{c}_{1_y} = 0, \qquad (3.45)$$

where

$$\dot{c}_{1_x} = \left(h_f \sin \delta_1 + c \cos \delta_1\right)\dot{\delta}_1 - \left(h_f \sin \delta_2 - c \cos \delta_2\right)\dot{\delta}_2,$$
$$\dot{c}_{1_y} = \left(h_f \cos \delta_2 + c \sin \delta_2\right)\dot{\delta}_2 - \left(h_f \cos \delta_1 - c \sin \delta_1\right)\dot{\delta}_1. \qquad (3.46)$$

Since Eq. (3.45) only depends on δ_1, δ_2 and $\dot{\delta}_1, \dot{\delta}_2$, we require only partial derivatives of $\dot{c}_1$ with respect to these variables for $\tilde{\boldsymbol{x}} = \tilde{\boldsymbol{x}}_0 \equiv \boldsymbol{0}$. Due to $\dot{c}_{1_x}\big|_* = \dot{c}_{1_y}\big|_* \equiv 0$ and $c_{1_x}\big|_* = 0$, this results in

$$\frac{\partial \dot{c}_1}{\partial \delta_1}\bigg|_* = 2\left(\frac{\partial c_{1_x}}{\partial \delta_1}\dot{c}_{1_x} + c_{1_x}\frac{\partial \dot{c}_{1_x}}{\partial \delta_1} + \frac{\partial c_{1_y}}{\partial \delta_1}\dot{c}_{1_y} + c_{1_y}\frac{\partial \dot{c}_{1_y}}{\partial \delta_1}\right)\bigg|_*$$

$$\equiv 2\left(c_{1_y}\frac{\partial \dot{c}_{1_y}}{\partial \delta_1}\right)\bigg|_* = 2[c_{1_y}\left(h_f \sin \delta_1 + c \cos \delta_1\right)\dot{\delta}_1]\big|_* \qquad (3.47)$$

$$= 0,$$

$$
\begin{aligned}
\left.\frac{\partial \dot{c}_1}{\partial \dot{\delta}_1}\right|_* &= 2\left.\left(c_{1_x}\frac{\partial \dot{c}_{1_x}}{\partial \dot{\delta}_1} + c_{1_y}\frac{\partial \dot{c}_{1_y}}{\partial \dot{\delta}_1}\right)\right|_* \\
&= 2[c_{1_x}\left(h_f \sin \delta_1 + c\cos \delta_1\right) - c_{1_y}\left(h_f \cos \delta_1 - c\sin \delta_1\right)]\big|_* \\
&\equiv -2c_{1_y}\big|_* \, h_f \equiv -2bh_f,
\end{aligned}
$$
(3.48)

where $c_{1_y}\big|_* = 2a - 2c \equiv b$ has been used again. Similarly, we find the derivatives with respect to δ_2, $\dot{\delta}_2$ as

$$
\left.\frac{\partial \dot{c}_1}{\partial \delta_2}\right|_* = 0, \quad \left.\frac{\partial \dot{c}_1}{\partial \dot{\delta}_2}\right|_* = 2bh_f.
$$
(3.49)

With analogous results for the rear axle constraint $\dot{c}_2 = 0$, the 2×10 Jacobian of the velocity constraints reads as

$$
\dot{\boldsymbol{C}}_{\tilde{\boldsymbol{x}}}\big|_* = \begin{bmatrix} 0 & \cdots & 0 & -2bh_f & 2bh_f & 0 & 0 \\ 0 & \cdots & 0 & 0 & 0 & 2bh_r & -2bh_r \end{bmatrix}.
$$
(3.50)

By concatenating submatrices (3.38), (3.44) and (3.50), we may construct $\tilde{\boldsymbol{T}}$ according to Eq. (3.26) as

$$
\tilde{\boldsymbol{T}} =
$$

$$
\left[\begin{array}{ccccccccc}
1/2 & 1/2 & 0 & 0 & \cdots & & & & 0 \\
0 & 0 & 1/2 & 1/2 & 0 & \cdots & & & 0 \\
0 & \cdots & & 0 & 1 & 0 & \cdots & & 0 \\
0 & \cdots & & 0 & 1 & 0 & \cdots & & 0 \\
0 & \cdots & & & 0 & 1/2 & 1/2 & 0 & 0 \\
0 & \cdots & & & 0 & 0 & 1/2 & 1/2 \\
\hline
-2bh_f & 2bh_f & 0 & 0 & 0 & \cdots & & & 0 \\
0 & 0 & 2bh_r & -2bh_r & 0 & \cdots & & & 0 \\
\hline
0 & \cdots & & & 0 & -2bh_f & 2bh_f & 0 & 0 \\
0 & \cdots & & & 0 & 0 & 0 & 2bh_r & -2bh_r
\end{array}\right].
$$
(3.51)

The first six columns of its inverse results in the reduced inverse $\widetilde{\mathcal{T}}$ as

$$\widetilde{T}^{-1} =$$

$$
\left[
\begin{array}{ccccccc|ccc|cc}
1 & 0 & \cdots & & & & 0 & -1/4bh_f & 0 & 0 & 0 \\
1 & 0 & \cdots & & & & 0 & 1/4bh_f & 0 & 0 & 0 \\
0 & 1 & 0 & \cdots & & & 0 & 0 & 1/4bh_r & 0 & 0 \\
0 & 1 & 0 & \cdots & & & 0 & 0 & -1/4bh_r & 0 & 0 \\
0 & 0 & 1 & 0 & \cdots & & 0 & 0 & 0 & 0 & 0 \\
0 & \cdots & & 0 & 1 & 0 & 0 & 0 & 0 & 0 & 0 \\
0 & \cdots & & & 0 & 1 & 0 & 0 & 0 & -1/4bh_f & 0 \\
0 & \cdots & & & 0 & 1 & 0 & 0 & 0 & 1/4bh_f & 0 \\
0 & \cdots & & & & 0 & 1 & 0 & 0 & 0 & 1/4bh_r \\
0 & \cdots & & & & 0 & 1 & 0 & 0 & 0 & -1/4bh_r \\
\end{array}
\right]
$$

$$\underbrace{}_{\widetilde{\mathcal{T}}}$$

$$(3.52)$$

defining the relation (3.27) between pseudo-state vector (3.33) and state vector (3.37). Finding the linear state-space representation (3.32) of the system from (3.29) is then straightforward and it may serve as basis for control design.

To demonstrate the state reduction procedure by using the projection matrices (3.51) and $\widetilde{\mathcal{T}}$ from (3.52), we may calculate Eqns. (3.26) and (3.27). For the sake of simplicity, let us investigate only the displacement of the front wheels and extract the decoupled equations one and seven from Eq. (3.26), i.e.,

$$
\begin{bmatrix} 1/2 & 1/2 \\ -2bh_f & 2bh_f \end{bmatrix}
\begin{bmatrix} \delta_1 \\ \delta_2 \end{bmatrix}
=
\begin{bmatrix} \delta_f \\ 0 \end{bmatrix},
\tag{3.53}
$$

with the expected result $\delta_1 = \delta_2 = \delta_f$ for $\tilde{x}_0 = 0$. Similarly, extraction of the first two equations from Eq. (3.27) gives

$$
\begin{bmatrix} 1 \\ 1 \end{bmatrix} \delta_f =
\begin{bmatrix} \delta_1 \\ \delta_2 \end{bmatrix}
\tag{3.54}
$$

resulting again in $\delta_1 = \delta_2 = \delta_f$. Analogous result may be obtained for the rear axle as well, i.e., $\delta_3 = \delta_4 = \delta_r$. Although this may look trivial here, the proposed procedure would also work for a non-trivial linearisation point of interest $\tilde{\boldsymbol{x}}_0 \neq \boldsymbol{0}$ resulting in $\delta_1 \neq \delta_2$, e.g., for a curved reference trajectory as might be required for a gain-scheduling control approach applied to a parameter-varying system.

3.4.2 Validation of the Linearised Model

Here only the validity of the linearisation method shall be checked. Therefore, step response simulations are performed for the uncontrolled nonlinear and linearised models with the model parameters listed in Table 5.1. Constant traction forces $F_{x_2} = -F_{x_1} = F_{x_4} = -F_{x_3} = 500\,\text{N}$ are applied as step inputs at $t = 0$ and state variables are obtained as response. The simulations are performed for different vehicle speeds, where the results can be seen in Fig. 3.2.

Obviously the linearised model follows its nonlinear counterpart with a reasonable accuracy regarding the results from a vehicle dynamics point of view. However, there are differences as well: Generally the linearised model works with lower errors in case of low vehicle speeds. This behaviour follows from the strongly nonlinear tyre models where at higher speeds the sideslip angles and hence the tyre forces and torques increase and the linearisation becomes less accurate.

Although the main goal here is not the assessment of the vehicle dynamics, we may conclude some basic properties of the vehicle model. Most importantly, the passive vehicle is stable at all investigated vehicle speeds despite the applied traction forces. The yaw response of the vehicle is reasonably fast, especially in case of $v = 80\,\text{km/h}$, where it exceeds its steady-state response within 0.2 s. A typical result is observable from the comparisons of results for $v = 30\,\text{km/h}$ and $v = 80\,\text{km/h}$, namely that the sideslip angle β in Fig. 3.2c changes its sign as the vehicle exceeds its tangent speed. We may also observe that the magnitudes of the state variables become lower as the vehicle speed increases. This is a consequence of the increasing tyre forces and torques while the traction forces remain the same in the different simulation cases.

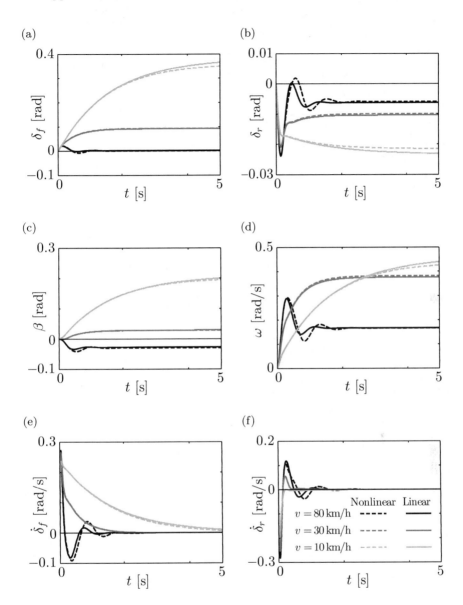

Figure 3.2: Comparison of simulation results for (a) front and (b) rear
            steering angle, (c) sideslip angle, (d) yaw rate, and (e) front
            and (f) rear steering angle speed

4 Control of the Differential Steering System

The vehicle model in Fig. 2.1 implies that the discussed differential steering concept is a steer-by-wire one. As an obvious consequence, the vehicle is unable to be steered correctly without using a control system. This chapter proposes control methods for that purpose.

As a first step, we have to process the driver's intention represented by some steering input $\delta_f^\star$ and transform it to either road wheel angles or vehicle dynamics reference values. For further discussions, let us separate high- and low-speed regimes by defining a transition speed v_t, which may be chosen almost arbitrarily, e.g., $v_t = 10\,\mathrm{km/h}$. In Section 4.1, a controller for higher speeds is developed based on the tracking of vehicle dynamics references, whereas in Section 4.2 road wheel angle control for lower speeds is discussed.

4.1 Full State Feedback Lateral Control for High-Speed Operation

Steering of conventional vehicles is enforced by turning the steering wheel and thus the wheel orientations δ_i through a steering mechanism. Common state-of-the-art steer-by-wire systems try to follow this by tracking control of the steering angle using the steering wheel angle as reference, see WANG ET AL. (2014) for example. Compared to such vehicles with dedicated steering actuators, here the cause-and-effect chain from the steering wheel angle through wheel orientations to resulting lateral and yaw motions is lost. Control of steering angles δ_f, δ_r, or solely aiming at control of steering linkages, is not advisable because it would unnecessarily split the control problem. Instead, the proposed control method makes the vehicle mimicking the behaviour of a reference vehicle. For this purpose, full state feedback control is used for lateral control of the vehicle at higher speeds, where the velocity-dependent control system tracks the behaviour of a simplified reference vehicle model.

Many studies apply reference model and yaw controller for superimposed vehicle dynamics control, see SAKAI ET AL. (1999) for example. However, by taking into account the considerations discussed in the preceding paragraph, exclusive control of the lateral and yaw dynamics of the vehicle is chosen similarly to DOMINGUEZ-GARCIA ET AL. (2004), instead of superimposed control. For high speed $v \geq v_t$, therefore, we have to calculate reference values

$$\boldsymbol{r}^* = \begin{bmatrix} \beta_{\text{ref}} & \omega_{\text{ref}} \end{bmatrix}^{\text{T}} \tag{4.1}$$

for sideslip angle β and yaw rate ω based on the steering input $\delta_f^\star$. The aim of the differential steering controller is then to follow these reference values $\boldsymbol{r}^*$ as close as possible.

4.1.1 Closed-Loop System

For control design, we need to extend the linear plant model (3.32) by system output $\hat{\boldsymbol{y}}$ as

$$\dot{\boldsymbol{x}} = \boldsymbol{A}\boldsymbol{x} + \boldsymbol{B}\boldsymbol{u}, \ \hat{\boldsymbol{y}} = \boldsymbol{C}\boldsymbol{x}. \tag{4.2}$$

The output matrix $\boldsymbol{C}$ maps β and ω as part of the state vector $\boldsymbol{x}$ to outputs $\hat{\boldsymbol{y}} = \begin{bmatrix} \beta & \omega \end{bmatrix}^{\text{T}} \in \mathbb{R}^2$, i.e.,

$$\boldsymbol{C} = \begin{bmatrix} 0 & 0 & 1 & 0 & 0 & 0 \\ 0 & 0 & 0 & 1 & 0 & 0 \end{bmatrix}. \tag{4.3}$$

According to LEVINE (2011b, p. 9/91), state feedback may be obtained as $\boldsymbol{u} = -\boldsymbol{K}\boldsymbol{x}$ with feedback gain matrix $\boldsymbol{K}$. However, a typically used positive definite $\boldsymbol{K}$ would drive $\boldsymbol{x}$ towards $\boldsymbol{0}$, which is not desirable here, because in our case we would like to set β and ω to prescribed reference values (4.1). This can be achieved by introducing a proper reference input $\hat{\boldsymbol{r}}_*$ and adapting the control input to $\boldsymbol{u} = \hat{\boldsymbol{r}}_* - \boldsymbol{K}\boldsymbol{x}$, which turns the regulator into a tracking controller where $\boldsymbol{x}$ is assumed to be known from ideal sensor readings.

Let us define the reference input as $\hat{\boldsymbol{r}}_* = \boldsymbol{F}\boldsymbol{r}^*$ based on Eq. (4.1) where $\boldsymbol{F}$ is a feedforward gain matrix. Combining these definitions and substituting them into Eq. (4.2) results in the control input

$$\boldsymbol{u} = \boldsymbol{F}\boldsymbol{r}^* - \boldsymbol{K}\boldsymbol{x} \tag{4.4}$$

and closed-loop dynamics

$$\dot{x} = (A - BK)\,x + BFr^*. \tag{4.5}$$

In the following, the feedback gain K is derived by using the LQ-principle, the feedforward gain F is obtained from the steady-state behaviour of the system, and the reference information r^* is deduced from a reference model.

4.1.2 Feedback Gain Calculation Using the LQ-Principle

In case of MIMO systems it is not obvious how to choose K in Eq. (4.4) or (4.5), respectively. A possible approach is to find the 'best' feedback gain according to some criteria, where a widely used method for this is the linear quadratic regulator (LQR) with certain advantages: it ensures the stability of the closed-loop system and it is easy to design and implement (LEVINE, 2011a, p. 17/1). Therefore, it is chosen in the following, although it may not be optimal for tracking control problems. By applying the infinite horizon LQR, finding the 'best' K is related to a cost function

$$J\,(u) = \int_{0}^{\infty} \left(x^{\mathrm{T}} Q x + u^{\mathrm{T}} R u \right) \mathrm{d}t, \tag{4.6}$$

where Q and R are positive semi-definite and positive definite weighting matrices for state vector x and input vector u, respectively. In order to solve the LQR design problem, we look for a particular K that minimises (4.6) when substituted into the closed-loop system (4.5). Particularly, it may be obtained as

$$K = R^{-1} B^{\mathrm{T}} S, \tag{4.7}$$

where S is the solution of the Riccati-equation

$$A^{\mathrm{T}} S A - S A R^{-1} B^{\mathrm{T}} S + Q = 0, \tag{4.8}$$

see OGATA (2010, pp. 793–795).

In the subsequent application, the choice of the input weight as $R = \mathrm{diag}\,\{r_1,\ldots,r_4\}$ with $r_1 = r_2$, $r_3 = r_4$ accounts for the symmetrical arrangement of left and right sides of the vehicle. The state weight is defined as $Q = \mathrm{diag}\,\{q_1,\ldots,q_6\}$ where the selection of q_i at first has to balance the influence of the different state variables. Since here the aim is to control β and ω as third and fourth coordinates of state vector (3.37) predominantly, q_3 and q_4 are selected to be higher than the other four weights, see Table 5.2. The specific choice of numbers is typically based on experience; for an optimal choice, the method in Chapter 6 or the concept of NGUYEN AND BESTLE (2007) may be used for example.

4.1.3 Feedforward Gain Calculation

Tracking of the reference input r^* by a proper feedforward gain matrix F in (4.4) should be performed with zero steady-state error. Alternatively, this may be achieved by integral action in the control loop according to POLMANS AND STRACKE (2014). However, the integral action may cause problems like overshoot and longer settling time (LI ET AL., 2014, p. 10), whereas fast response is more important here than complete elimination of the steady-state error as the driver is part of the vehicle control for compensation. Although integral control is more robust against disturbances, this is also not required here. In real disturbance scenarios like side wind or uneven roads, the driver will act as an integral controller to compensate the disturbances just as in case of conventional vehicles.

According to these considerations, integral control is not the right choice here. Instead, the previously designed LQR is extended to an LQ-tracker according to LEVINE (2011a, p. 25/19). For this, let us define the steady-state output as $\hat{y}_\infty = C x_\infty$ with the steady state vector resulting from the condition $\dot{x} = 0$ in Eq. (4.5) as

$$x_\infty = -\left(A - BK\right)^{-1} BFr^*. \tag{4.9}$$

Error-free steady state requires the output of Eq. (4.2) to be identical with the reference input (4.1), i.e., $\hat{y}_\infty \overset{!}{=} r^*$ or

$$-C\left(A - BK\right)^{-1} BFr^* \overset{!}{=} r^*. \tag{4.10}$$

The equation is satisfied for arbitrary reference inputs r^* if

$$-C\left(A - BK\right)^{-1} BF = I \tag{4.11}$$

resulting in

$$F = -\left[C\left(A - BK\right)^{-1} B\right]^{+}, \tag{4.12}$$

where $^+$ denotes the Moore–Penrose inverse. By using such a gain F, we can properly pre-amplify r^* and thus reach zero steady-state error. In other words, F is the matrix of the inverse steady-state gains of the closed-loop system (4.5).

4.1.4 Reference Model

Also in case of steer-by-wire, the driver (or any autonomous driving assistance system) has to apply a steering input $\delta_f^\star$, e.g., by a steering wheel to account for the actual driving situation. For the concept described in Section 4.1.1, this information then has to be transformed into $\boldsymbol{r}^* = \begin{bmatrix} \beta_{\text{ref}} & \omega_{\text{ref}} \end{bmatrix}^{\text{T}}$ acting as reference for control input (4.4). This transformation $\delta_f^\star \longrightarrow \boldsymbol{r}^*$ can be achieved by using a reference model as shown in Fig. 4.1, which can be chosen almost arbitrarily. Any model may be used which is able to provide sideslip angle β and yaw rate ω.

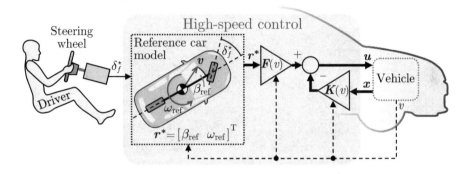

Figure 4.1: Driver interaction with vehicle through reference model

Due to its simplicity, here the well-known linear single track vehicle model (RIEKERT AND SCHUNCK, 1940) with only front wheel steering $\delta_f^\star$ is applied as reference model. As a practical result, the steering controller may mimic the behaviour of any desired conventional vehicle and thus a familiar driving feel. In state-space form the equations of the single track model read as

$$\dot{\boldsymbol{r}}^* = \boldsymbol{A}_r \boldsymbol{r}^* + \boldsymbol{B}_r \delta_f^\star \tag{4.13}$$

for $\boldsymbol{r}^* = \begin{bmatrix} \beta_{\text{ref}} & \omega_{\text{ref}} \end{bmatrix}^{\text{T}}$ where the wheel steering reference $\delta_f^\star$ may be related to the steering wheel position by multiplying it with a steering transmission ratio. The coefficient matrices $\boldsymbol{A}_r$ and $\boldsymbol{B}_r$ may be taken from RILL (2012, pp. 198–190) as

$$\boldsymbol{A}_r = \begin{bmatrix} -\dfrac{C_{F_{\alpha_f}} + C_{F_{\alpha_r}}}{mv} & \dfrac{C_{F_{\alpha_r}} l_r - C_{F_{\alpha_r}} l_f}{mv^2} - 1 \\ \dfrac{C_{F_{\alpha_r}} l_r - C_{F_{\alpha_r}} l_f}{J_{b_z}} & -\dfrac{C_{F_{\alpha_r}} l_r^2 + C_{F_{\alpha_r}} l_f^2}{J_{b_z} v} \end{bmatrix}, \quad \boldsymbol{B}_r = \begin{bmatrix} \dfrac{C_{F_{\alpha_f}}}{mv} \\ \dfrac{C_{F_{\alpha_f}} l_f}{J_{b_z}} \end{bmatrix}, \tag{4.14}$$

where $C_{F_{\alpha_f}} = 2C_{F_\alpha}(F_{z_f}/2)$, $C_{F_{\alpha_r}} = 2C_{F_\alpha}(F_{z_r}/2)$ are the cornering stiffness parameters of the front and rear axles, respectively. Selection of the reference model parameters is up to the designer's judgement. In this case, parameter values are taken from the vehicle model, see Table 5.1.

4.1.5 Gain Scheduling Extension

In reality, the coefficient matrices in the linearised equations (3.32) of the vehicle model depend on the speed v of the vehicle, i.e., $\boldsymbol{A} = \boldsymbol{A}(v)$, $\boldsymbol{B} = \boldsymbol{B}(v)$. In order to account for this, the vehicle model may be regarded as a gain scheduling system depending on vehicle speed, and the controller parameters in Eqns. (4.4), (4.12) and $\boldsymbol{K}$ in Section 4.1.2 should be treated also as functions of speed, i.e.,

$$\boldsymbol{u} = \boldsymbol{F}(v)\,\boldsymbol{r}^* - \boldsymbol{K}(v)\,\boldsymbol{x}, \qquad (4.15)$$

where $\boldsymbol{K}(v)$ and $\boldsymbol{F}(v)$ follow from repeating the calculations in Sections 4.1.2 and 4.1.3 for various speeds v, respectively. This is not a true parameter-varying extension, but a simple gain scheduling system as characterised by SHAMMA AND ATHANS (1991), where the derivative of the vehicle speed with respect to time is not considered.

4.2 Angle Tracking Controller for Low-Speed Operation

For low speeds, the steering is mainly characterised by the kinematics of the steering linkage, because lateral forces are rather low and wheel motion follows the wheel orientation (HARRER AND PFEFFER, 2017, p. 29). We may conclude from this property that it is advantageous not to apply the control concept used for higher speeds, but to realise a behaviour similar to the conventional Ackermann steering by applying a control action such that the steering linkage(s) are forced into the desired direction(s). Here, the low-speed regime is characterised as $v \leq v_t = 10\,\text{km/h}$ as a rather arbitrary choice, which in practical applications depends on the desired character of the car to be designed.

4.2.1 PI Control Rule

A straightforward solution for such an angle tracking controller is to apply a simple PI angle control to the front axle only and keep the rear wheels straight, as used in most papers dealing with steer-by-wire systems. The controller output is the traction force

$$F_{x_2} = P\left(\delta_f^\star - \delta_f\right) + I \int \left(\delta_f^\star - \delta_f\right) \mathrm{d}t \qquad (4.16)$$

on the right front wheel resulting in $F_{x_1} = -F_{x_2}$ for the left wheel in order to enforce pure steering without longitudinal acceleration. The coefficients P and I are the proportional and integral gains of the controller. For simplification, the rear axle is kept passive and rather straight $(\delta_r \approx 0)$ by taking a high value for the rear axle stiffness c_r and setting $F_{x_3} = F_{x_4} = 0$. For the low-speed regime, the complete control input finally is

$$\boldsymbol{u} = \begin{bmatrix} -F_{x_2} & F_{x_2} & 0 & 0 \end{bmatrix}^\mathrm{T}. \qquad (4.17)$$

The choice of c_r is a compromise between allowing enough turning for high speed, but keeping the vehicle predominantly front-steered at all speeds. The corresponding parameter selection is part of the optimisation concept in Chapter 6. It should be mentioned that stiffness c_r allows some passive turning of the rear wheels also at low speed which is taken into consideration by the nonlinear vehicle model (2.36) used for simulation.

4.2.2 Control Design with Root Locus Method

For the control design of the low-speed controller, let us start from the Laplace transform of the linearised system (3.32) with substituted control input (4.17) yielding the transfer matrix

$$\boldsymbol{G}_{F_{x_2} \to \boldsymbol{x}} = (s\boldsymbol{I} - \boldsymbol{A})^{-1}\boldsymbol{B} \begin{bmatrix} -1 & 1 & 0 & 0 \end{bmatrix}^\mathrm{T}. \qquad (4.18)$$

Due to the definition (3.37) of the state vector, the first element of the transfer matrix (4.18) is the transfer function $G_{F_{x_2} \to \delta_f}(s)$ between traction force F_{x_2} and front steering angle δ_f. Combined with the PI controller (4.16) represented by the control transfer function $C_L(s) = P + I/s$, the transfer function of the complete closed loop in Fig. 4.2 follows as

$$G_{\delta_f^\star \to \delta_f} = \frac{C_L G_{F_{x_2} \to \delta_f}}{1 + C_L G_{F_{x_2} \to \delta_f}}. \qquad (4.19)$$

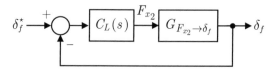

Figure 4.2: Low-speed control loop

The aim of the control design procedure is to find proper P and I coefficients ensuring the desired behaviour of the closed loop (4.19).

In principle, the coefficients P and I of control (4.16) may be directly chosen as design variables. However, randomly selected P and I values likely result in poor tracking performance or even numerical problems. A further problem is that the feasible lower and upper limits for P and I may depend on mechanical parameters. In order to avoid such difficulties, we should not chose P and I directly, but apply a procedure as proposed in the following.

Let us approximate the transfer function $G_{F_{x_2} \to \delta_f}(s)$ by a first order system

$$\widetilde{G}_{F_{x_2} \to \delta_f}(s) := \frac{A}{\tau s + 1} \approx G_{F_{x_2} \to \delta_f}(s), \qquad (4.20)$$

where $A = \lim_{s \to 0} G_{F_{x_2} \to \delta_f}(s)$ is the static gain and $\tau = -1/\mathrm{Re}(s_0)$ the time constant associated with the dominant pole s_0 of $G_{F_{x_2} \to \delta_f}$. Thus, A and τ are deduced from given system information (4.18). Combined with the control transfer function C_L, the approximate transfer function of the closed loop (4.19) follows as

$$\widetilde{G}_{\delta_f^\star \to \delta_f} = \frac{C_L \widetilde{G}_{F_{x_2} \to \delta_f}}{1 + C_L \widetilde{G}_{F_{x_2} \to \delta_f}} = \frac{A(P + I/s)}{(\tau s + 1) + A(P + I/s)}$$

$$= \frac{A(Ps + I)}{\tau s^2 + (AP + 1)s + AI}. \qquad (4.21)$$

Solution of its characteristic equation $\tau s^2 + (AP + 1)s + AI = 0$ results in two distinct poles $s_1 = -1/\tau$ and $s_2 = -AP/\tau$, where the dominant one (s_1) may be cancelled by choosing $I = P/\tau$. This can be seen by substituting this in the closed-loop dynamics (4.21) resulting in

$$\widetilde{G}_{\delta_f^\star \to \delta_f} \approx \frac{AP(s + 1/\tau)}{\tau s^2 + (AP + 1)s + AP/\tau} \equiv \frac{AP(s + 1/\tau)}{\tau(s + 1/\tau)(s + AP/\tau)}$$

$$\equiv \frac{AP}{\tau s + AP}. \qquad (4.22)$$

The remaining pole $s_2 = -AP/\tau$ may be placed anywhere along the root locus curve. In order to place it to a specific position $s_2 = -1/\tau_d$, we may calculate P as

$$P = \frac{1}{A}\frac{\tau}{\tau_d}, \tag{4.23}$$

where A and τ are given by the approximation (4.20), and τ_d is the desired time constant of the closed loop. For the parameters in Table 5.1, we obtain $A = 1.099 \cdot 10^{-3}$ and $\tau = 9.69\,\mathrm{s}$ resulting in Fig. 4.3.

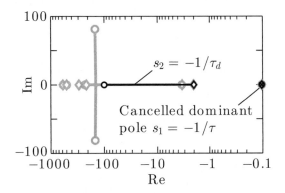

Figure 4.3: Root locus curves of the original ($G_{\delta_f^* \to \delta_f}$, light) and approximated ($\widetilde{G}_{\delta_f^* \to \delta_f}$, dark) transfer functions for root locus ranging from $\tau_d = 0.01\,\mathrm{s}$ (circle) to $\tau_d = 0.5\,\mathrm{s}$ (diamond)

5 Simulations and Steering Characterisation

In order to assess the fundamental behaviour and characteristics of the discussed differential steering concept, we have to perform simulations by combining the vehicle model and the control algorithms discussed in Chapters 2 and 4 into a joint simulation framework. In order to make the simulation results comparable to results reported in the literature, the applied simulation scenarios follow the typical vehicle dynamics test standards. Besides the simulations, a formal characterisation of the steering performance is introduced for optimisation purposes. For the latter, the same simulation framework is utilised, although we have to define different simulation scenarios designed specifically for optimisation. The chapter is organised as follows: Section 5.1 introduces the applied simulation framework and model parameters. Section 5.2 discusses some fundamental simulation results, while Section 5.3 covers the steering performance characterisation.

5.1 Simulation Framework

The simulation framework combines the nonlinear vehicle model (2.36) with the speed-dependent control algorithms (4.15), (4.16) and (4.17) according to Fig. 5.1. The parameters of the vehicle model and control algorithms are provided in Tables 5.1 and 5.2, respectively. Applied vehicle parameters rely on experimental data and are derived from LUNDAHL ET AL. (2011) except the tyre-related parameters which are taken from PACEJKA (2006) and RILL (2012). The applied parameter set represents a typical C-segment passenger car. The maximal traction force $F_x^{\max}$ limiting $|F_{x_i}| \leq F_x^{\max}$, $i = 1 \ldots 4$, is derived from PEROVIC (2012) representing a realisation of an in-wheel motor capable for this traction force.

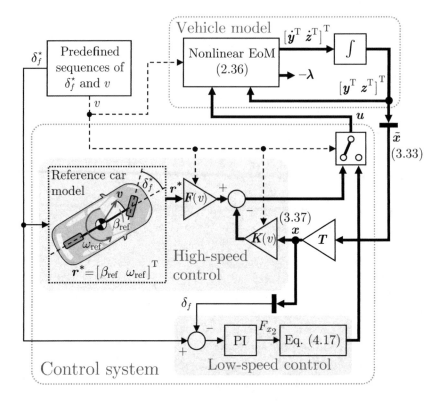

Figure 5.1: Simulation model for test manoeuvres

5.2 Simulation Studies

In order to assess the properties of the discussed steering concept, some simulation experiments are presented including step response, steady-state cornering, double lane change and tracking manoeuvres at low speeds.

5.2.1 Step Steer Simulation

The step response test is fundamental for assessing the lateral dynamics of a vehicle. During the test, a sudden angle change is applied to the steering wheel represented by front wheel reference angle $\delta_f^\star$, and state variables like ω or β are captured until a certain level of steady state is reached in the lateral acceleration. Dynamic properties like rise time or overshoot may be concluded from this manoeuvre.

Table 5.1: Vehicle parameters: (a) geometry, (b) kingpin stiffness and damping, (c) mass and inertia, and (d) tyre-related parameters

(a)

Symbol	Value	Unit
l_f	1.03	m
l_r	1.55	m
a	0.703	m
b	1.23	m
h_f, h_r	0.14	m
r_f	0.078	m
r_r	-0.075	m
t_f	-0.01	m
t_r	0.045	m

(b)

Symbol	Value	Unit
c_f	1	Nm/rad
c_r	2000	Nm/rad
d	80	Nms/rad

(c)

Symbol	Value	Unit
m_b	1150	kg
m_w	40	kg
J_{b_z}	2500	kgm^2
J_w	1.84	kgm^2
J_{w_r}	2.01	kgm^2

(d)

Symbol	Value	Unit
l_c	0.16	m
w_c	0.195	m
h	0.52	m
F_{z_0}	4000	N
C_y	1.3	-
E_y	-1	-
C_z	2.3	-
E_z	-2	-
c_1	8.16	-
c_2	2.33	-
c_3	0.18	-
c_4	0.3	-
c_9	0.3	-
c_{10}	0	-

Table 5.2: Control parameters

Symbol	Value	Unit
Q	diag $\left\{10^{-3}, 10^{-3}, 5 \cdot 10^{6}, 10^{5}, 10^{-3}, 10^{-3}\right\}$	-
R	diag $\left\{10^{-3}, 10^{-3}, 10^{-3}, 10^{-3}\right\}$	-
τ_d	0.5	s
$F_x^{\max}$	2000	N

Here, the standardised version according to ISO 7401 is used with $\delta_f^\star$ as steering input, where the increase of the steering input $\dot{\delta}_f^\star = 0.64\,\text{rad/s}$ is comparable to an angular velocity of $500\,°/\text{s}$ of a conventional vehicle's steering wheel assuming a steering ratio of 13.6:1. The vehicle speed is $v = 80\,\text{km/h}$ resulting in a steady-state lateral acceleration of $4\,\text{m/s}^2$, as can be seen in Fig. 5.2b.

The most important result of this test is the rise time T_ω of the yaw rate ω indicating how quickly the vehicle responds to the steering input. According to Fig. 5.2c, T_ω is defined as the time span between the points where the steering input $\delta_f^\star$ reaches half of its maximum, i.e., $\delta_f^\star = \delta_{f_{\max}}^\star/2$,

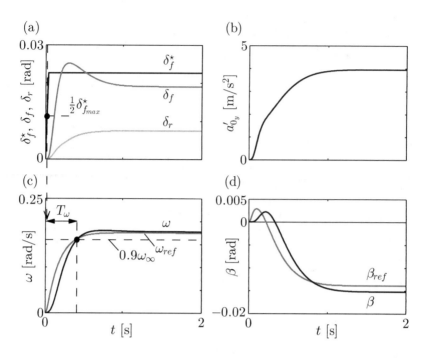

Figure 5.2: Step steer manoeuvre: (a) steering angles, (b) lateral acceleration, (c) yaw response and (d) sideslip angle

and where ω reaches 90% of the steady-state value ω_∞. Here we obtain
$T_\omega \approx 0.4\,\mathrm{s}$, whereas typical rise times for passenger cars are in the range
0.1–$0.3\,\mathrm{s}$ as may be found from experimental data (CHELI ET AL., 2007).
Thus, we may conclude that the rise time here is slightly higher than usual.
This follows, however, mainly from the behaviour of the reference model
in Section 4.1.4, because actually the controller follows the references with
only $0.1\,\mathrm{s}$ delay as shown in Figs. 5.2c and 5.2d. Thus, the rise time might
be reduced by using a more agile reference model, for example a model
representing a sports car. A concept for designing a vehicle model close
to a prescribed dynamic behaviour may be found in BUSCH AND BESTLE
(2011).

For assessing control quality, we may introduce the steady-state error

$$e_{\omega_\infty} = \left| \frac{(\omega_{\mathrm{ref}_\infty} - \omega_\infty)}{\omega_{\mathrm{ref}_\infty}} \right| \tag{5.1}$$

and overshoot

$$o_\omega = \left| \frac{(\omega_{\max} - \omega_\infty)}{\omega_\infty} \right|. \tag{5.2}$$

One may conclude from Fig. 5.2c that ω reaches its steady state with a slight overshoot of about $o_\omega = 1\%$ and with $e_{\omega_\infty} = 2\%$ steady-state error. In case of β, the steady-state error with an analogous definition is $e_{\beta_\infty} = 9\%$.

5.2.2 Steady-State Cornering

The steady-state cornering test is performed according to ISO 4138, where the vehicle runs on a circular path with radius $R = 100\,\text{m}$. Its speed is increased only very slowly to maintain a quasi-steady behaviour, and the steering angle is adapted in order to keep the path. The test runs as long as the vehicle is able to keep track, which in our case is up to a lateral acceleration of about $a'_{0_y} \approx 8.5\,\text{m/s}^2$, see Fig. 5.3.

Further, self-steering properties of the vehicle might be concluded from this test. According to Fig. 5.3, the vehicle shows understeering behaviour while $a'_{0_y} \leq 5.3\,\text{m/s}^2$ with an initial self-steering gradient of about

$$\left. \frac{\partial \delta^\star_f}{\partial a'_{0_y}} \right|_{a'_{0_y}=0} \approx 0.03\, \frac{^\circ}{\text{m/s}^2} \tag{5.3}$$

and oversteering beyond.

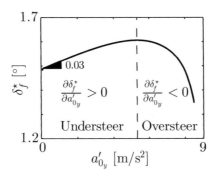

Figure 5.3: Necessary steering during steady-state cornering

Certain characteristic values of vehicle speed may be obtained as well. For example, from Fig. 5.4a we get $\beta = 0$ for $a'_{0_y} \approx 2.3\,\text{m/s}^2$ which, according to the centrifugal relation $a'_{0_y} = v'^2_{0_x}/R \approx v^2/R$, corresponds to tangential speed $v \approx \sqrt{Ra'_{0_y}} = 54.5\,\text{km/h}$.

The sideslip angle gain $\beta/\delta_f^\star$ and yaw rate gain $\omega/\delta_f^\star$ in Fig. 5.5 are unbounded due to the oversteering characteristic. Figure 5.5b shows that the yaw rate gain goes to infinity at $a_{0_y}' \approx 8.5\,\mathrm{m/s^2}$ associated with critical speed $v_C \approx \sqrt{Ra_{0_y}'} \approx 105\,\mathrm{km/h}$. These results (except the oversteering behaviour) are typical for passenger cars (PAUWELUSSEN, 2014).

The tracking performance of the controller can be estimated based on Fig. 5.4. The controller is able to track both β_{ref} and ω_{ref} with negligible error up to $a_{0_y}' \approx 4.5\,\mathrm{m/s^2}$. As the vehicle speed and hence lateral acceleration increase, the vehicle behaves more and more nonlinearly increasing the steady-state error up to $e_{\beta_\infty} = 137\,\%$ and $e_{\omega_\infty} = 39\,\%$, respectively, when reaching the critical speed.

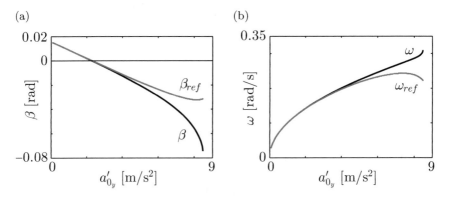

Figure 5.4: Tracking performance during steady-state cornering:
(a) sideslip angle and (b) yaw rate

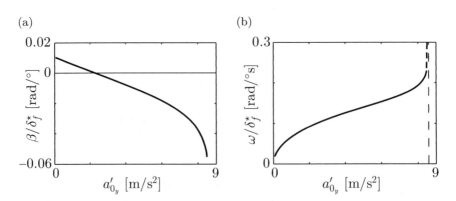

Figure 5.5: Sideslip angle gain (a) and yaw rate gain (b) during steady-state cornering

5.2.3 Double Lane Change

The double lane change manoeuvre imitates the situation of avoiding an impact with an obstacle suddenly appearing in front of the vehicle. The test procedure applied here follows the ISO 3888-2 standard which defines that the minimum speed is $v = 60\,\text{km/h}$ (also used here) and the road surface is dry ($\mu \approx 1$). The test is passed if the track envelope remains untouched and the vehicle keeps its stability.

Figure 5.6 shows the test track and the main simulation result. The trajectory of the vehicle CoG and the area swept by the vehicle body during the manoeuvre do not touch the track envelope, and thus the test is passed. The corresponding controlled variables are shown in Fig. 5.7, where the yaw rate ω follows its reference ω_{ref} with a reasonable delay and accuracy. Tracking of the sideslip angle reference β_{ref} is a little worse, mainly because at certain points the vehicle encounters its sliding limits where the linear controller cannot provide optimal performance.

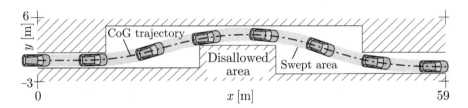

Figure 5.6: Double lane change test track and vehicle's trajectory for $v = 60\,\text{km/h}$

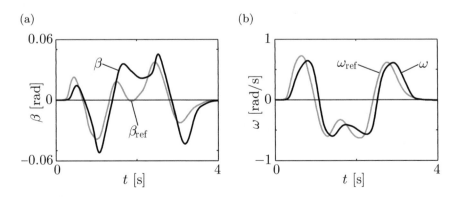

Figure 5.7: Reference tracking of (a) sideslip angle and (b) yaw rate during double lane change

We may also investigate the tyre forces shown in Fig. 5.8. The obtained radii of the $\mu F_{z_i}^{min}$ and $\mu F_{z_i}^{max}$, $i = 1 \ldots 4$, circles relative to that of μF_{z_0} indicate that the vertical load of the wheels largely varies due to the lateral load transfer. Both the longitudinal and lateral tyre forces of the front wheels encounter their limits, although there is no clear sign of saturation, i.e., the vehicle remains controllable during the manoeuvre.

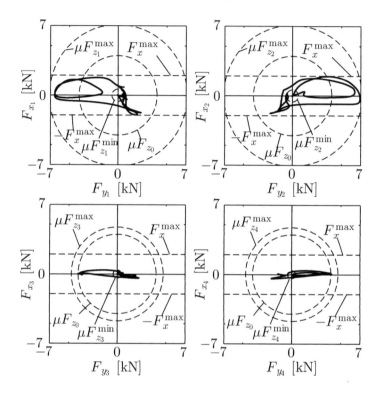

Figure 5.8: Tyre forces during double lane change

5.2.4 Low-Speed Manoeuvring

In order to investigate a parking-like manoeuvre at lower speeds, angle tracking simulations are performed. End-to-end steering movements are applied using a triangle wave reference signal $\delta_f^\star$, see Fig. 5.9a. The controller is obviously able to track the reference, and the tracking performance even improves as the vehicle speed increases.

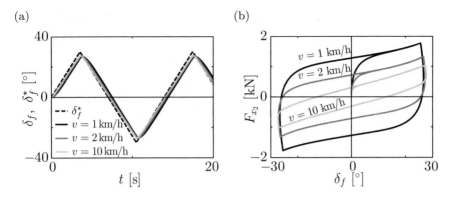

Figure 5.9: Low-speed manoeuvring: (a) angle tracking and (b) traction
force hysteresis

A further question is, if the system is able to generate the necessary
traction forces being shown in Fig. 5.9b as function of the steering angle
(force hysteresis loop). Obviously the traction force does not exceed the
actuator limit $F_{x_i}^{\mathrm{max}} = 2000\,\mathrm{N}$ and the force hysteresis loop collapses
gradually as the vehicle speed is increased.

5.3 Steering Performance Characterisation

Based on the discussed simulation results, we now need to deduce criteria evaluating the steering performance for the optimisation procedure
introduced in Chapter 6. For this purpose, one may prefer to assess the
performance by test manoeuvres similarly to the preceding simulations
rather than by artificial objectives. In the time domain, the dynamic
performance is assessed by using step response investigations, where an
alternative formulation is discussed in the frequency domain. Additionally, steady-state cornering and low-speed manoeuvring capabilities are
discussed. All of these cases are associated with well-defined objective values. By combining these characteristics, we may obtain a comprehensive
picture of the vehicle's steering performance from multiple perspectives.

5.3.1 Dynamic Performance in the Time Domain

Usually, the dynamic performance in the time domain is tested by step response according to ISO 7401 as discussed already in Section 5.2.1. Various results may be obtained from such tests, but the most important values with respect to steering dynamics are the rise times T_β and T_ω of β and ω, respectively, as they show how agile the vehicle reacts on the steering input. As already discussed, the rise times may be calculated as time spans between the time where the steering input $\delta_f^\star$ reaches half of its maximum $\delta_{f_{max}}^\star = \max_t \delta_f^\star(t)$ and the time where $\beta = 0.9\beta_\infty$ or $\omega = 0.9\omega_\infty$, respectively, where β_∞ and ω_∞ are steady-state values of β and ω, see Fig. 5.10a and 5.10b.

As the different cases in Fig. 5.10 demonstrate, such an investigation of rise times may not be robust enough for optimisation. While $T_\beta = 0.856$ s and $T_\omega = 0.398$ s represent the steering performance rather well in Fig. 5.10a, the parameter setting in Fig. 5.10b with comparable results $T_\beta = 0.692$ s and $T_\omega = 0.405$ s is obviously worse due to the oscillatory behaviour. Moreover, some parameter variations may even result in such a slow dynamics that the steady state is not reached during simulation time, and thus the rise times cannot even be calculated as shown in Fig. 5.10c. Therefore, an alternative evaluation criteria is required.

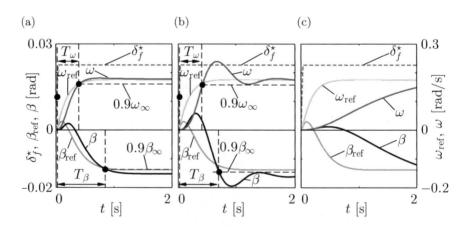

Figure 5.10: Step responses for different state weights **Q**: (a) regular, (b) oscillatory and (c) slow behaviour (scale for δ and β is shown on the left, that of ω on the right)

In our case, the control method in Fig. 4.1 suggests such an alternative approach for performance assessment. As the reference model already represents the desired dynamics of the vehicle and the control strategy aims at tracking β_{ref} and ω_{ref} with associated values for T_β and T_ω, design objectives may be based on the quality of tracking. This may be characterised by the root mean square (RMS) tracking errors

$$e_\beta = \sqrt{\frac{1}{t_{\max}} \int_0^{t_{\max}} (\beta - \beta_{\text{ref}})^2 dt}, \quad e_\omega = \sqrt{\frac{1}{t_{\max}} \int_0^{t_{\max}} (\omega - \omega_{\text{ref}})^2 dt}, \quad (5.4)$$

where $t_{\max}$ is the considered simulation time (e.g., $t_{\max} = 2\,\text{s}$ in Fig 5.10). In the above example, tracking errors $e_\beta = 0.002$, $e_\omega = 0.015$ for Fig. 5.10a and $e_\beta = 0.005$, $e_\omega = 0.039$ for Fig. 5.10b now show a significant performance difference correlating with human perception. Parameter values are taken from Tables 5.1 and 5.2, except $\boldsymbol{Q} = \text{diag}\left\{5 \cdot 10^{-5}, 5 \cdot 10^{-5}, 8 \cdot 10^{-7}, 10^{-5}, 10^{-4}, 1.2 \cdot 10^{-4}\right\}$ for Fig. 5.10b and $\boldsymbol{Q} = \text{diag}\left\{5 \cdot 10^6, 5 \cdot 10^6, 5 \cdot 10^{-6}, 10^{-6}, 10^7, 1.2 \cdot 10^7\right\}$ for Fig. 5.10c, respectively.

5.3.2 Tracking Performance in the Frequency Domain

Investigations in the time domain are rather time-consuming, being an adverse property for optimisation. By characterising the tracking performance in the frequency domain, some of the difficulties discussed in Fig. 5.10 may be avoided and the dynamic behaviour is assessed in a wider range than in the time domain.

For the frequency-domain description, let $G_{\beta_{\text{ref}} \to \beta}$ and $G_{\omega_{\text{ref}} \to \omega}$ denote the transfer functions from reference values β_{ref} and ω_{ref} to actual states β and ω, respectively. Then the ideal tracking behaviour would be given by

$$G_{\beta_{\text{ref}} \to \beta} = 1 \quad \text{and} \quad G_{\omega_{\text{ref}} \to \omega} = 1. \tag{5.5}$$

These transfer functions may be obtained from a Laplace transformation of the closed loop system (4.5) resulting in

$$\mathscr{L}\left\{\dot{\boldsymbol{x}}\right\} \equiv s\mathscr{L}\left\{\boldsymbol{x}\right\} = (\boldsymbol{A} - \boldsymbol{BK})\,\mathscr{L}\left\{\boldsymbol{x}\right\} + \boldsymbol{BF}\mathscr{L}\{\boldsymbol{r}^*\}. \tag{5.6}$$

By denoting Laplace transforms of $\boldsymbol{x}$ and $\boldsymbol{r}^*$ as $\boldsymbol{x}_{\mathscr{L}}(s) = \mathscr{L}\left\{\boldsymbol{x}(t)\right\}$ and $\boldsymbol{r}^*_{\mathscr{L}}(s) = \mathscr{L}\left\{\boldsymbol{r}^*(t)\right\}$, we get

$$\boldsymbol{x}_{\mathscr{L}} = (s\boldsymbol{I} - \boldsymbol{A} + \boldsymbol{BK})^{-1}\,\boldsymbol{BF}\boldsymbol{r}^*_{\mathscr{L}} =: \boldsymbol{G}_{\boldsymbol{r}^* \to \boldsymbol{x}}\,\boldsymbol{r}^*_{\mathscr{L}}. \tag{5.7}$$

Due to definitions $r^* = \begin{bmatrix} \beta_{\text{ref}} & \omega_{\text{ref}} \end{bmatrix}^{\text{T}}$ and $x = \begin{bmatrix} \delta_f & \delta_r & \beta & \omega & \dot{\delta}_f & \dot{\delta}_r \end{bmatrix}^{\text{T}}$, the elements (3,1) and (4,2) of the 6×2 transfer matrix $G_{r^* \to x}$ provide the transfer functions of interest

$$G_{\beta_{\text{ref}} \to \beta}(s) \equiv G_{r_1^* \to x_3}(s) \quad \text{and} \quad G_{\omega_{\text{ref}} \to \omega}(s) \equiv G_{r_2^* \to x_4}(s). \qquad (5.8)$$

Substituting $s = j\widehat{\omega} = j2\pi f$ results in conditions for ideal tracking behaviour (5.5) given by magnitude one and phase zero responses, i.e.,

$$\begin{aligned} |G_{\beta_{\text{ref}} \to \beta}(\hat{f})| &\overset{!}{=} 1, \ \angle G_{\beta_{\text{ref}} \to \beta}(\hat{f}) \overset{!}{=} 0, \\ |G_{\omega_{\text{ref}} \to \omega}(\hat{f})| &\overset{!}{=} 1, \ \angle G_{\omega_{\text{ref}} \to \omega}(\hat{f}) \overset{!}{=} 0 \ \forall \hat{f}, \end{aligned} \qquad (5.9)$$

where the dimensionless logarithmic frequency $\hat{f} = \log_{10}(f/1\,\text{Hz})$ is used in Fig. 5.11 and all further calculations (MATTA ET AL., 2011).

Tracking performance may then be characterised by calculating deviations of the actual frequency responses from this ideal behaviour (5.9). It is sufficient to consider only magnitude responses, because in case of lowpass-like characteristics, phase responses will tend towards zero automatically

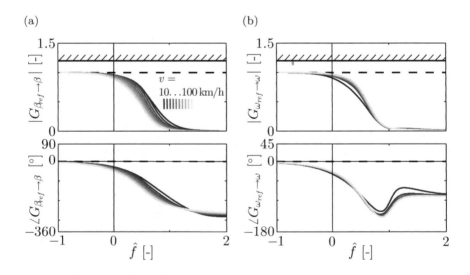

Figure 5.11: Dynamic performance criteria based on transfer functions (a) $G_{\beta_{\text{ref}} \to \beta}$ and (b) $G_{\omega_{\text{ref}} \to \omega}$ with desired characteristics (dashed) and upper bounds (hatched)

if the magnitude responses tend towards the unit gain, see Bode's gain-phase relation (BODE, 1940). Further, we may conclude the time-domain behaviour from frequency-domain results. For example, the higher the cutoff frequency, the better the vehicle performs in step or slalom tests.

Since responses are frequency dependent as shown in Fig. 5.11, the tracking performance may be characterised by the mean deviations from the unit gain as

$$\overline{\Delta A_\beta} = \frac{1}{\hat{f}_{max} - \hat{f}_{min}} \int_{\hat{f}_{min}}^{\hat{f}_{max}} \left|1 - \left|G_{\beta_{ref} \to \beta}(\hat{f})\right|\right| d\hat{f},$$

$$\overline{\Delta A_\omega} = \frac{1}{\hat{f}_{max} - \hat{f}_{min}} \int_{\hat{f}_{min}}^{\hat{f}_{max}} \left|1 - \left|G_{\omega_{ref} \to \omega}(\hat{f})\right|\right| d\hat{f},$$

(5.10)

where $\hat{f}_{min} = \log_{10}(0.1\,\mathrm{Hz}/1\,\mathrm{Hz}) = -1$ and $\hat{f}_{max} = \log_{10}(100\,\mathrm{Hz}/1\,\mathrm{Hz}) = 2$. From Fig. 5.11 it becomes further obvious that the system and thus the responses depend on vehicle speed v. In order to take into account a specific speed range $[v_{min}, v_{max}]$, the means

$$\overline{\overline{\Delta A_\beta}} = \frac{1}{v_{max} - v_{min}} \int_{v_{min}}^{v_{max}} \overline{\Delta A_\beta}(v)\, dv \quad \text{and}$$

$$\overline{\overline{\Delta A_\omega}} = \frac{1}{v_{max} - v_{min}} \int_{v_{min}}^{v_{max}} \overline{\Delta A_\omega}(v)\, dv$$

(5.11)

may be considered as representative performance measures, where, e.g., $v_{min} = 10\,\mathrm{km/h}$ and $v_{max} = 100\,\mathrm{km/h}$.

High gains may lead to uncomfortable or even dangerous situations by causing, e.g., driver-induced oscillations. This can be avoided by taking into account upper bounds

$$G_\beta^{max} := \max_{\hat{f},v} \left|G_{\beta_{ref} \to \beta}(\hat{f}, v)\right| \leq A_\beta^{max} \quad \text{and}$$

$$G_\omega^{max} := \max_{\hat{f},v} \left|G_{\omega_{ref} \to \omega}(\hat{f}, v)\right| \leq A_\omega^{max},$$

(5.12)

where the limit values in Fig. 5.11 are chosen as $A_\beta^{max} = A_\omega^{max} = 1.2$.

5.3.3 Steady-State Cornering Performance

As discussed in Section 5.2.2, the steady-state cornering performance may be investigated according to ISO 4138 by driving the vehicle on a circular path with increasing speed and adjusted steering angle until it is not able to keep the path anymore. Alternatively, the vehicle speed may be kept constant while the steering angle is increased slowly resulting in a spiral trajectory, see Fig. 5.12a. In the latter case, the necessity of an additional steering controller or driver model is eliminated, which results in a more robust procedure for optimisation. Therefore, it shall be used here.

The cornering performance may then be characterised by the maximally achieved lateral acceleration

$$a_{0_y}^{\prime\,\mathrm{max}} = \max_t a_{0_y}^{\prime}\left(\delta_f^{\star}(t)\right), \quad \text{where} \quad \delta_f^{\star}(t) = \dot{\delta}_f^{\star} t \tag{5.13}$$

is related to a prescribed steering velocity $\dot{\delta}_f^{\star} = \mathrm{const.}$, see Fig. 5.12b. In order to avoid lengthy simulations or calculation difficulties, the simulation terminates if any of the following conditions occurs:

- the traction force exceeds the friction circle limit;
- any wheel lifts off the road surface ($F_{z_i} \leq 0$);
- all drive motors reach the maximal traction force F_x^{max}, i.e., $|F_{x_i}| \geq F_x^{\mathrm{max}} \,\forall i$;
- a steering limit δ_{max} is reached, i.e., $\max\{\delta_f^{\star}, |\delta_f|, |\delta_r|\} \geq \delta_{\mathrm{max}}$;

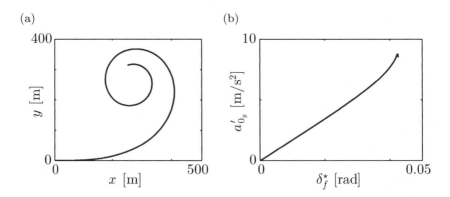

Figure 5.12: Quasi-steady cornering with constant speed $v = 80\,\mathrm{km/h}$ resulting in (a) spiral trajectory and (b) corresponding lateral acceleration

- the sideslip angle reaches some maximum $\beta_{\max}$, i.e., $|\beta| \geq \beta_{\max}$;
- the gradient of the lateral acceleration becomes negative and the acceleration has decreased sufficiently, i.e., $da'_{0_y}/d\delta^\star_f < 0 \wedge a'_{0_y} < 0.9 \max_t a'_{0_y}$, since β and ω slowly drift away from β_{ref} and ω_{ref} as a'_{0_y} increases and the lateral tyre forces saturate.

Applied parameter values are $\delta_{\max} = 0.6\,\mathrm{rad}$, $\dot{\delta}^\star_f = 7.5 \cdot 10^{-4}\,\mathrm{rad/s}$, $F^{\max}_x = 2000\,\mathrm{N}$ and $\beta_{\max} = 0.3\,\mathrm{rad/s}$, where $\delta_{\max}$ is selected as average maximal steering angle of a typical passenger car, $\dot{\delta}^\star_f$ as a value being low enough to keep the quasi-steady behaviour, $\beta_{\max}$ is selected rather arbitrarily, and $F^{\max}_x$ stems from the maximal traction force capability of the in-wheel motors, see Table 5.2.

5.3.4 Low-Speed Manoeuvring Performance

The low-speed manoeuvring performance may be characterised by investigating the tracking performance of the low-speed controller (4.16) during a manoeuvre which represents the typical lock-to-lock steering movements during parking and low-speed manoeuvring, see Fig. 5.13a. During this manoeuvre, the controller tries to track the triangular wave reference steering angle $\delta^\star_f$ requiring purposefully selected PI-gains.

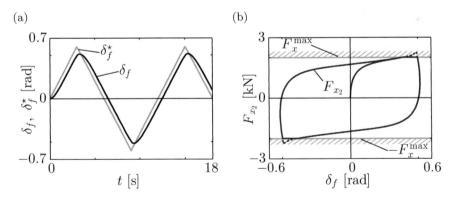

Figure 5.13: Low-speed manoeuvring for $v = 1\,\mathrm{km/h}$ and $\tau_d = 0.5\,\mathrm{s}$: (a) reference tracking and (b) traction force hysteresis limited by $F^{\max}_x$

Since the applied manoeuvre here is more demanding than that in Section 5.2.4, the tracking performance may degrade as the maximally achievable traction force F_x^{max} limits the control input, see Fig. 5.13b. Deviations may then be characterised by the mean tracking error as low-speed manoeuvring performance

$$\overline{\Delta\delta}_f = \frac{1}{t_{\mathrm{max}}} \int_0^{t_{\mathrm{max}}} \left| \delta_f^\star (t) - \delta_f (t) \right| \mathrm{d}t, \tag{5.14}$$

where simulation time is $t_{\mathrm{max}} = 18\,\mathrm{s}$ and the reference steering velocity is chosen as $\dot{\delta}_f^\star = (1/5)\,\mathrm{sgn}\,(\cos{(\pi t/6)})$.

6 Multi-Objective Steering Performance Optimisation

As the application of pure differential steering in passenger cars does not exist yet, we do not know its applicability in general and do not have well-established 'rules of thumb' for design like those we have for conventional steering systems, neither for the mechanical part nor for its control. In order to explore the general suitability and find proper designs, e.g., as starting point for prototype production and field tests, optimisation may be applied similarly to ZHAO ET AL. (2012) where the steering dynamics and road feel of a DDAS are optimised by adapting various mechanical quantities, and to ZHAO ET AL. (2018, 2019) where the vehicle model and objectives of ZHAO ET AL. (2012) are modified by the inclusion of suspension ride comfort aspects. ZHAO ET AL. (2011) present an optimisation method for DDAS in order to improve the steering feel, the steering portability and the stability. LI ET AL. (2015) optimise the control performance of a DDAS by adjusting control rules of an associated fuzzy controller. However, steering optimisation should not be restricted to either mechanical or control parameters, but they can be adapted simultaneously with a joint optimisation procedure to exploit the full optimisation potential of the strongly coupled mechanical and control parameters. For example, BUSCH AND BESTLE (2014) present such an approach for robust design of a more conventional all-wheel-steering vehicle.

For proof of general applicability of the differential steering concept as the only steering component and finding suitable designs, this chapter presents a comprehensive multi-objective optimisation concept based on the design objectives and constraints introduced in Section 5.3. Due to the strong coupling between mechanical and control parts, the set of design variables includes both mechanical and control system parameters resulting in a joint optimisation problem, which is then solved by a response surface aided multi-objective genetic algorithm. After discussing the design parametrisation, a sensitivity analysis is performed to find the most influential parameters. Finally, the multi-objective optimisation problem is formulated and solved, and some specific optimisation results are discussed.

6.1 Design Parametrisation

Design parametrisation is a key-driver deciding on failure or success of optimisation. The goal is to select the most influential and reasonably adjustable parameters, where any parameter of the vehicle model or control system may serve as a design variable.

Selectable parameters of the control system are the design weights of the high-speed control system in Eq. (4.6), whereas τ_d may be preselected according to Section 6.2.1. Since control parameters q_i and r_j are typically varied within multiple orders of magnitude, logarithmic weights $\hat{q}_i = \log_{10} q_i$, $i = 1 \ldots 6$, and $\hat{r}_j = \log_{10} r_j$, $j = 1, 3$, are introduced as design parameters instead. Mechanical parameters of the steering system in Fig. 2.1 are the trail dimensions (t_f, t_r), scrub radii (r_f, r_r), kingpin stiffnesses (c_f, c_r) and damping d, as well as steering linkage dimensions a, c, h_f and h_r. By summarising them, the vector of potential design variables may be preliminarily defined as

$$\boldsymbol{p} := \left[\boldsymbol{p}_c^{\mathrm{T}} \mid \boldsymbol{p}_m^{\mathrm{T}}\right]^{\mathrm{T}} \in \mathbb{R}^{19}$$

$$= \left[\hat{q}_1 \ldots \hat{q}_6 \ \ \hat{r}_1 \ \ \hat{r}_3 \mid t_f \ \ t_r \ \ r_f \ \ r_r \ \ c_f \ \ c_r \ \ d \ \ a \ \ c \ \ h_f \ \ h_r\right]^{\mathrm{T}},$$
$$\tag{6.1}$$

where $\boldsymbol{p}_c \in \mathbb{R}^8$ and $\boldsymbol{p}_m \in \mathbb{R}^{11}$ are control and mechanical parameters, respectively.

Reasonable lower and upper bounds $\boldsymbol{p}^l := \left[\boldsymbol{p}_c^{l\,\mathrm{T}} \mid \boldsymbol{p}_m^{l\,\mathrm{T}}\right]^{\mathrm{T}}$ and $\boldsymbol{p}^u := \left[\boldsymbol{p}_c^{u\,\mathrm{T}} \mid \boldsymbol{p}_m^{u\,\mathrm{T}}\right]^{\mathrm{T}}$ are selected according to Table 6.1. For control parameters, they are chosen by the consideration that state weights $\hat{q}_i$ should be higher than input weights $\hat{r}_j$ to prioritise tracking performance over control effort. Bounds for the mechanical parameters are chosen by common sense, taking into account that the resulting geometries should fit into a typical passenger car. For example, trails and scrub radii are bounded in the ± 0.1 m range in order to constrain the volume swept by the wheels, except that a negative scrub radius r_f is not advisable for the front axle. Torsional spring stiffnesses are chosen such that the front steering linkage is relaxed compared to the rear to ensure a front-wheel dominant steering characteristic. The general rule is that optimisation should not drive parameter values to their bounds, else they should be released, except they are hard limits. Comparison with results in Table 6.3 shows that bounds have no influence on the final optimal designs.

Table 6.1: Lower and upper limit values of (a) control and (b) mechanical parameters (given in SI-units)

(a)

	$\hat{q}_1$	$\hat{q}_2$	$\hat{q}_3$	$\hat{q}_4$	$\hat{q}_5$	$\hat{q}_6$	$\hat{r}_1$	$\hat{r}_3$
$\boldsymbol{p}_c^l$	-3	-3	-3	-3	-3	-3	-3	-3
$\boldsymbol{p}_c^u$	8	8	10	8	8	8	2	2

(b)

	t_f	t_r	r_f	r_r	c_f	c_r	d	a	c	h_f	h_r
$\boldsymbol{p}_m^l$	-0.1	-0.1	0.05	-0.1	0	500	20	0.65	-0.05	0.1	0.1
$\boldsymbol{p}_m^u$	0.1	0.1	0.1	0.1	100	10^4	200	0.75	0.15	0.2	0.2

6.2 Sensitivity Studies

In advance of optimisation, sensitivity analyses should be performed in order to reveal possible correlations between performance measures, determine the most influential parameters, and find a reasonable value for τ_d.

6.2.1 Preselection of Control Parameter τ_d

In order to find a proper value for the time constant τ_d determining the low-speed manoeuvring dynamics (4.23), a preliminary design of experiments (DoE) is applied where 100 designs are generated by space-filling optimised Latin hypercube sampling (oLHS) maximising the minimum distance between the sampled designs (MORRIS AND MITCHELL, 1995). As the low-speed control performance only depends on τ_d and the mechanical parameters $\boldsymbol{p}_m$, the oLHS is taken concerning only the latter s.t. $\boldsymbol{p}_m^l \leq \boldsymbol{p}_m \leq \boldsymbol{p}_m^u$. Effects of parameter variations of mechanical parameters $\boldsymbol{p}_m$ and τ_d on the low-speed performance measure (5.14) may be seen in Fig. 6.1 as variety of functions $\overline{\Delta\delta}_f = \overline{\Delta\delta}_f(\tau_d)$.

Obviously, lower values of τ_d generally yield lower tracking errors, regardless to variations of mechanical parameters $\boldsymbol{p}_m$. This suggests that τ_d should not be considered as a design variable, but chosen as low as possible. However, low τ_d may result in aggressive control interventions which should be avoided. Considering that the controller saturates and only marginal performance improvements are obtained when $\tau_d \lesssim 0.1\,\mathrm{s}$, we may preselect $\tau_d = 0.1\,\mathrm{s}$ for further investigations.

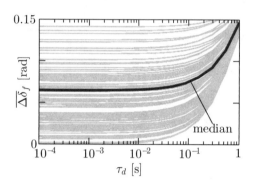

Figure 6.1: Characteristics of low-speed tracking error $\overline{\Delta\delta}_f$ as function of τ_d for various $\boldsymbol{p}_m$-settings

6.2.2 Identification of the Most Influential Parameters

In order to find the most influential parameters, a global DoE is performed. A sample set of $n_s = 2000$ designs is generated for the complete design vector (6.1) by oLHS in the range $\boldsymbol{p}^l \leq \boldsymbol{p} \leq \boldsymbol{p}^u$. The generated designs are evaluated by simulation of the controlled, nonlinear vehicle model (Fig. 5.1) and calculation of criteria (5.11)–(5.14) resulting in an evaluated sample set as a basis for further investigations. Some of these designs resulted in extreme values, which are considered as outliers. Table 6.2a shows some selected criterion limits and the percentage of designs exceeding these limits, which are removed from the sample set. Finally, 1709 designs remain for further studies.

The DoE further showed that maximum values G_β^{max} and G_ω^{max} of transfer functions (5.12) are distributed rather unevenly which may lead to difficulties in the forthcoming calculations, e.g., training of regression models. In order to ease processing, logarithmic transformations

Table 6.2: Limits and removed percentages of the sample set (a) and most significant design parameters (b)

			$\overline{\overline{\Delta A_\beta}}$	$\overline{\overline{\Delta A_\omega}}$	$a_{0_y}'^{\,\max}$	$\overline{\Delta\delta}_f$	$G_\beta^{\max}$	$G_\omega^{\max}$
(a)	Upper limit		0.5	0.5	10	0.4	10	10
	Removed %		6.1	6.8	0	6.6	4.2	6.5
(b)		1	$\hat{q}_3$	$\hat{q}_4$	t_f	t_f	r_r	r_r
		2	$\hat{q}_5$	t_f	t_r	c	c_r	c_r
		3	c_r	$\hat{r}_1$	r_r	r_f	t_r	t_r
	Order of significance	4	t_r	$\hat{q}_5$	c_r	h_f	$\hat{q}_3$	t_f
		5	$\hat{r}_1$	$\hat{q}_3$	r_f		t_f	$\hat{q}_4$
		6	$\hat{q}_6$	r_r	$\hat{r}_1$		$\hat{r}_1$	$\hat{r}_1$
		7	d	$\hat{r}_3$	$\hat{r}_3$		$\hat{r}_3$	$\hat{q}_3$
		8	$\hat{r}_3$	c_r	$\hat{q}_5$		$\hat{q}_4$	$\hat{r}_3$
		9	t_f	t_r			$\hat{q}_5$	c_f
		10	$\hat{q}_4$	$\hat{q}_6$			$\hat{q}_6$	d
		11	r_r	$\hat{q}_1$			d	$\hat{q}_6$
		12	$\hat{q}_1$				$\hat{q}_1$	$\hat{q}_5$
		13					c_f	c
		14						$\hat{q}_1$

$\widehat{G}_j^{\max} := \log_{10} G_j^{\max}$, $j \in \{\beta, \omega\}$, are applied resulting in logarithmised variants of constraints (5.12), i.e.,

$$\widehat{G}_j^{\max} \leq \widehat{A}_j^{\max}, \quad \text{where } \widehat{A}_j^{\max} := \log_{10} A_j^{\max} = 0.079. \tag{6.2}$$

Next, the most influential parameter subset on each of the criteria $\xi(\boldsymbol{p})$ representing one of the functions (5.11)–(5.14) may be selected based on the results of the global DoE. The selection is done by iteratively sorting the parameters with respect to their explanatory influence, see Fig. 6.2a. For this procedure, let us investigate the explanatory power of a subset of design parameters (6.1) summarised in a reduced design vector $\boldsymbol{\chi} := \begin{bmatrix} p_{j_1} & p_{j_2} & \dots \end{bmatrix}^{\mathrm{T}}$, $\{j_1, j_2, \dots\} \subset \{1, \dots, n_p\}$, by training regressive metamodels $\xi(\boldsymbol{p})$ and $\hat{\xi}(\boldsymbol{\chi})$ on the labelled sets $(\boldsymbol{p}^{(k)}, \xi^{(k)})$, $\xi^{(k)} = \xi(\boldsymbol{p}^{(k)})$, and $(\boldsymbol{\chi}^{(k)}, \hat{\xi}^{(k)})$, $\hat{\xi}^{(k)} = \hat{\xi}(\boldsymbol{\chi}^{(k)})$, $k \in \{1, \dots, n_s\}$, with the regression conditions

$$\hat{\xi}(\boldsymbol{\chi}^{(k)}) \overset{!}{=} \xi^{(k)}, \quad \boldsymbol{\chi}^{(k)} = \begin{bmatrix} p_{j_1}^{(k)} & p_{j_2}^{(k)} & \dots \end{bmatrix}^{\mathrm{T}}. \tag{6.3}$$

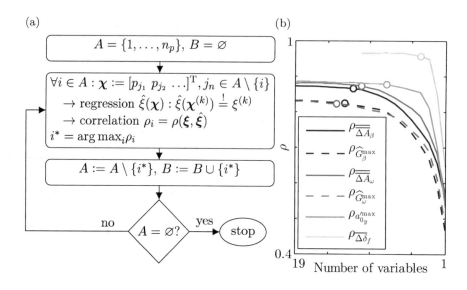

Figure 6.2: Recursive parameter sorting w.r.t. increasing influence (a) based on regression quality (b) (circles denote first significant variables, respectively)

The explanatory power of $\boldsymbol{\chi}$ may then be assessed through regression quality represented by the empirical Pearson's correlation coefficient

$$\rho(\boldsymbol{\xi}, \hat{\boldsymbol{\xi}}) = \frac{\sum\limits_{k=1}^{n_s} \left(\xi^{(k)} - \bar{\xi} \right) \left(\hat{\xi}^{(k)} - \bar{\hat{\xi}} \right)}{\sqrt{\sum\limits_{k=1}^{n_s} \left(\xi^{(k)} - \bar{\xi} \right)^2} \sqrt{\sum\limits_{k=1}^{n_s} \left(\hat{\xi}^{(k)} - \bar{\hat{\xi}} \right)^2}}, \tag{6.4}$$

where values close to one represent strong, while close to zero weak correlation, and $\bar{\xi} = \sum_{k=1}^{n_s} \xi^{(k)}/n_s$ is the mean, likewise $\bar{\hat{\xi}}$, too. The vector $\boldsymbol{\xi} = \left[\xi^{(1)} \ldots \xi^{(n_s)} \right]^{\mathrm{T}}$ summarises the exactly evaluated values on the basis of all n_p design parameters (6.1) and therefore represents the maximum of information contained in the dataset, while $\hat{\boldsymbol{\xi}} = \left[\hat{\xi}^{(1)} \ldots \hat{\xi}^{(n_s)} \right]^{\mathrm{T}}$ summarises the function values of an approximate metamodel $\hat{\xi}(\boldsymbol{\chi})$ trained on the reduced vector $\boldsymbol{\chi}$ of design variables.

Here, a regression tree with 100-fold bootstrap aggregation is applied as an approximate metamodel, as it is reported to be robust and flexible enough to deal with significant nonlinearities (WEHBI ET AL., 2017). This means that the parameter space is split into regions and a constant value is associated with each region such that the mean square of errors $\hat{\boldsymbol{\xi}} - \boldsymbol{\xi}$ is minimised. In order to avoid overfitting, the 'leafiness' of the tree is limited, i.e., the parameter space is partitioned in such a way that not less than 15 training data points are associated with each leaf. This procedure is repeated 100-times with bootstrapping and the results are aggregated into a smoothed metamodel by averaging the individual trees. A simplified example of the procedure is shown in Fig. 6.3 where the low-speed tracking error $\overline{\Delta\delta}_f$ is approximated with only the mechanical parameters c and t_f, i.e., $\xi = \overline{\Delta\delta}_f$ and $\boldsymbol{\chi} = \begin{bmatrix} c & t_f \end{bmatrix}^{\mathrm{T}}$.

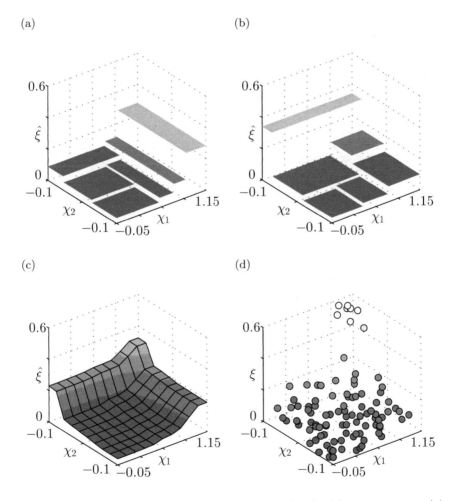

Figure 6.3: Regression tree responses of two individual bootstrappings (a), (b) of the aggregated tree (c) and original data (d)

By recursively removing the parameters with the least explanatory influence, i.e., the parameter whose removal results in the least decrease of (6.4), we remain with the most significant ones. The sorting is done over the parameter indices, starting from the complete index set A of left parameters and the empty set B of removed ones. The sorting procedure successively empties A and fills up B with parameter indices in increasing order of significance.

The sorting is performed for each of the critera (5.11)–(5.14) separately, see Fig. 6.2b. Even if all 19 parameters are used in the metamodel, i.e., $\hat{\xi}(\boldsymbol{\chi}) = \hat{\xi}(\boldsymbol{p})$, there will be approximation errors characterised by Pearson's correlation values $\rho(\boldsymbol{\xi}, \hat{\boldsymbol{\xi}}) < 1$. Subsequent removal of parameters further reduces the correlation resulting in monotonically decreasing curves. Circles show the limiting number of variables where the correlation value is still above 99% of the initial correlation value when using all design variables in the metamodel. The remaining, and thus significant, parameters are shown in Table 6.2b. Obviously $\overline{4\text{–}14}$ design variables are sufficient for each criterion where, e.g., $\overline{\Delta\delta}_f$ does not depend on control parameters and the significant parameter sets are different for each criterion.

The union of these parameters include 16 of the originally 19 design parameters. Thus, we may introduce a reduced parametrisation

$$\hat{\boldsymbol{p}} := \begin{bmatrix} \hat{\boldsymbol{p}}_c^{\mathrm{T}} \mid \hat{\boldsymbol{p}}_m^{\mathrm{T}} \end{bmatrix}^{\mathrm{T}} \in \mathbb{R}^{16}$$

$$= \begin{bmatrix} \hat{q}_1 & \hat{q}_3 & \dots & \hat{q}_6 & \hat{r}_1 & \hat{r}_3 \mid t_f & t_r & r_f & r_r & c_f & c_r & d & c & h_f \end{bmatrix}^{\mathrm{T}}$$

$$(6.5)$$

for optimisation, with corresponding lower and upper limits $\hat{\boldsymbol{p}}^l$ and $\hat{\boldsymbol{p}}^u$ taken from Table 6.1. For the excluded parameters we may use the mean of the corresponding lower and upper limit values given in Table 6.1 and keep them constant during optimisation, i.e.,

$$\begin{bmatrix} \hat{q}_2 & a & h_r \end{bmatrix}^{\mathrm{T}} = \begin{bmatrix} 2.5 & 0.7 & 0.15 \end{bmatrix}^{\mathrm{T}} = \mathrm{const.} \qquad (6.6)$$

Although it is obviously not possible to significantly reduce the overall number of design parameters, the reduced parameter sets of each individual criterion provide advantages anyway. As shown later in Section 6.3.2, the training of the response surface models (RSMs) may be largely simplified by using the individually reduced parameter sets improving the execution speed of the optimisation procedure.

6.3 Optimisation Strategy

The search for optimised designs will be performed by the multi-objective genetic algorithm NSGA-II requiring several thousands of design evaluations. Each evaluation of criteria (5.11)–(5.14) takes about 8–10 s by using an average desktop computer. Application of RSMs may speed up the search procedure (VENTER ET AL., 1996) which is why RSM-assistance is used here to solve the optimisation problem in a reasonable time.

6.3.1 Formulation of the Optimisation Problem

In order to formally set up the optimisation problem, we have to define design objectives and constraints. While $\overline{\Delta A}_\beta$, $\overline{\Delta A}_\omega$ and $a_{0_y}'^{\max}$ are related to the high-speed regime, $\overline{\Delta \delta}_f$ assesses the low-speed manoeuvring and may be considered as a separate operating mode. In addition, $\overline{\Delta \delta}_f$ does not depend on control parameters $\boldsymbol{p}_c$ as seen above. Therefore, it is straightforward to treat $\overline{\Delta \delta}_f$ not as a performance measure, but as a constraint. In order to minimise criteria (5.11) and maximise (5.13), we may formally write the vector of design objectives to be minimised as

$$\boldsymbol{o}(\hat{\boldsymbol{p}}) = \left[\overline{\Delta A}_\beta(\hat{\boldsymbol{p}}) \quad \overline{\Delta A}_\omega(\hat{\boldsymbol{p}}) \quad -a_{0_y}'^{\max}(\hat{\boldsymbol{p}})\right]^{\mathrm{T}} \to \min. \qquad (6.7)$$

The constraints (5.12) in logarithmic form (6.2) and the limitation of (5.14) by an upper bound result in inequality constraints

$$\boldsymbol{h}(\hat{\boldsymbol{p}}) = \begin{bmatrix} \max \left|\widehat{G}_{\beta_{\mathrm{ref}}\to\beta}(\hat{\boldsymbol{p}}; \hat{f}, v)\right| - \widehat{A}_\beta^{\max} \\ \max \left|\widehat{G}_{\omega_{\mathrm{ref}}\to\omega}(\hat{\boldsymbol{p}}; \hat{f}, v)\right| - \widehat{A}_\omega^{\max} \\ \overline{\Delta\delta}_f(\hat{\boldsymbol{p}}_m) - \overline{\Delta\delta}_f^{\max} \end{bmatrix} \leq \boldsymbol{0}, \qquad (6.8)$$

where the allowed mean tracking error is set as $\overline{\Delta\delta}_f^{\max} = 0.05\,\mathrm{rad}$.

Design vector (6.5), objectives (6.7) and constraints (6.8) result in a constrained optimisation problem. However, as a genetic algorithm is used for solution, we need to transform it to an unconstrained one by applying a penalty strategy realised by quadratic penalty terms with offset:

$$\min_{\hat{\boldsymbol{p}} \in \mathcal{P}} \left[\boldsymbol{o}\left(\hat{\boldsymbol{p}}\right) + \begin{bmatrix} 10 \\ 10 \\ 10 \end{bmatrix} \sum_{i=1}^{3} \left(\max\{0, h_i\} \right)^2 + \boldsymbol{\gamma}_0 \right] \tag{6.9}$$

$$\text{s.t. } \mathcal{P} := \left\{ \hat{\boldsymbol{p}} \in \mathbb{R}^{16} \,\middle|\, \hat{\boldsymbol{p}}^l \leq \hat{\boldsymbol{p}} \leq \hat{\boldsymbol{p}}^u \right\},$$

with offsets

$$\boldsymbol{\gamma}_0 = \begin{cases} \begin{bmatrix} 3 & 3 & 30 \end{bmatrix}^{\mathrm{T}} & \text{if } \exists i : h_i(\hat{\boldsymbol{p}}) > 0, \\ \boldsymbol{0} & \text{otherwise.} \end{cases} \tag{6.10}$$

This means that, if any of the constraints (6.8) is violated, the same quadratic penalty term will be added to all objectives (6.7), while the offsets (6.10) are defined such that the typical magnitude differences within $\boldsymbol{o}$ are taken into account to ensure a clear discrimination of the penalised values.

6.3.2 Optimisation Assistance by Response Surfaces

There is a variety of possibilities for defining response surfaces, like Kriging (CRESSIE, 1993) or radial basis functions (POWELL, 1987). Here, single hidden layer feedforward neural networks (LESHNO ET AL., 1993) are applied as RSMs, Fig. 6.4. The neural networks have 4–14 selected design parameters as inputs $\boldsymbol{\chi} \in \mathbb{R}^n$ depending on the number n of significant design parameters of each performance criterion found in Section 5.2, see Table 6.2b, and the same number of neurons in the hidden layer. The activation functions

$$f_h^{(i)} := \tanh\left(w_{h_0}^{(i)} + \boldsymbol{W}_h^{(i,:)} \boldsymbol{\chi} \right), \quad i = 1 \ldots n, \tag{6.11}$$

are hyperbolic tangent functions applied to the sum of input parameters weighted with the ith row of weight matrix $\boldsymbol{W}_h \in \mathbb{R}^{n \times n}$ and shifted by the ith coordinate of bias vector $\boldsymbol{w}_{h_0} \in \mathbb{R}^n$. The single output function

$$f_o := w_{o_0} + \boldsymbol{w}_o^{\mathrm{T}} \boldsymbol{f}_h \tag{6.12}$$

is a linear combination of the hidden layer outputs where w_{o_0} and $\boldsymbol{w}_o \in \mathbb{R}^n$ are the output bias and weights, respectively. The output of the neural network delivers the value of the response surface f_{RSM} and thus the estimate $\hat{\xi}$ as function of the input $\boldsymbol{\chi}$:

$$\begin{aligned} \hat{\xi} = f_{\mathrm{RSM}}(\boldsymbol{\chi}; \boldsymbol{w}_{h_0}, \boldsymbol{W}_h, w_{o_0}, \boldsymbol{w}_o) &:= w_{o_0} + \boldsymbol{w}_o^{\mathrm{T}} \boldsymbol{f}_h \\ &= w_{o_0} + \boldsymbol{w}_o^{\mathrm{T}} \tanh\left(\boldsymbol{w}_{h_0} + \boldsymbol{W}_h \boldsymbol{\chi} \right). \end{aligned} \tag{6.13}$$

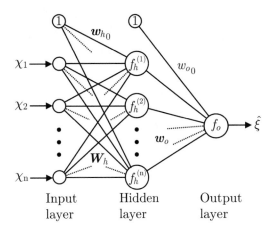

Figure 6.4: Single hidden layer neural network defining response surface
$\hat{\xi}(\boldsymbol{\chi})$

In order to mitigate the risk of overfitting, the neural network is trained
according to the following considerations. Typical back-propagation algorithms drive the prediction error

$$e_{\hat{\xi}} = \sum_{i=1}^{n_s} \left(\hat{\xi}(\boldsymbol{\chi}^{(i)}) - \xi(\boldsymbol{\chi}^{(i)}) \right)^2 \tag{6.14}$$

towards zero by adjusting the network weights. However, taking only error
(6.14) into account may lead to overfitting and loss of generalisation capabilities. A necessary condition of overfitting is the non-smooth response of
the network resulted from large network weights. This adverse tendency
can be mitigated by restricting the weights through regularisation which
means the extension of error (6.14) as

$$\tilde{e}_{\hat{\xi}} = \tilde{w}_1 e_{\hat{\xi}} + \tilde{w}_2 \sum_{i=1}^{n_w} w_i^2, \tag{6.15}$$

where $\tilde{w}_1$ and $\tilde{w}_2$ are purposefully chosen weights, and the second term
sums up all the n_w scalar network weights contained in $\boldsymbol{w}_{h_0}$, $\boldsymbol{W}_h$, w_{o_0}
and $\boldsymbol{w}_o$. If the network is trained s.t. error (6.15) is minimised, then large
network weights and thus overfitting can be avoided. However, it is a
question how to find a balance between $\tilde{w}_1$ and $\tilde{w}_2$. Here, Bayesian regularisation is applied, which is a statistical approach based on Bayes' rule to
find optimal values of $\tilde{w}_1$ and $\tilde{w}_2$, see DAN FORESEE AND HAGAN (1997).
Eventually, the neural network is trained by back-propagation extended

with Bayesian regularisation which is less prone to overfitting compared to pure gradient-based training methods without regularisation.

Although regression quality may be estimated by using only training data, this is usually misleading regarding the generalisation capabilities of the response surface. A more realistic picture may be obtained by cross-validation. Here, a 20-fold cross-validation is applied splitting dataset ξ of exact evaluations into 20 equal-sized subsets from which 19 are used for training and the remaining one for testing. The procedure is repeated 20 times always leaving out another subset for testing. The resulting $\tilde{\rho}_i$, $i = 1 \ldots 20$, correlation coefficients from each training-test cycle are then aggregated into the averaged correlation coefficient $\rho = \sum_{i=1}^{20} \tilde{\rho}_i/20$ representing the regression quality of each RSM, see Fig. 6.5 also showing summarised scatter plots of the 20 cross-validation cases. One may conclude from the corresponding averaged correlation coefficients ρ that predictions for $\widehat{G}_{\beta}^{\max}$ and $\widehat{G}_{\omega}^{\max}$ are particularly weak. However, we may consider instead whether the fulfilment of inequality constraints $\widehat{G}_{j}^{\max} \leq \widehat{A}_{j}^{\max}$, $j \in \{\beta, \omega\}$, is predicted correctly. The corresponding RSMs are still capable to predict this with 80.0% and 74.1% accuracy for β and ω, respectively, which is sufficient for optimisation.

We may investigate not only the regression quality, but also the execution time of the RSMs. A single evaluation of criteria (5.11)–(5.14) takes ≈ 10 s, while the same with the RSMs takes only 0.086 s which results in a speed-up by a factor of about 116. This property is utilised in Section 6.3.3 where an optimisation procedure based on RSMs is introduced and a detailed execution time analysis is discussed.

6.3.3 Optimisation Procedure

By using the RSMs, we may now set up and execute the optimisation procedure. An adaptive approach derived from REGIS AND SHOEMAKER (2004) is used, where the RSMs are updated iteratively according to Fig. 6.6. The procedure starts with the already evaluated sample set resulting from the DoE used for sensitivity studies in Section 6.2.2, here being utilised for a first RSM-training according to Section 6.3.2. Subsequently, an optimum search is applied on these initial RSMs by using a non-dominated sorting genetic algorithm (NSGA-II) from DEB (2001), since it is gradient-free, global and applicable to multi-objective problems. Within NSGA-II, an initial population consisting of 200 designs is generated by oLHS, and then a search runs for up to 100 generations

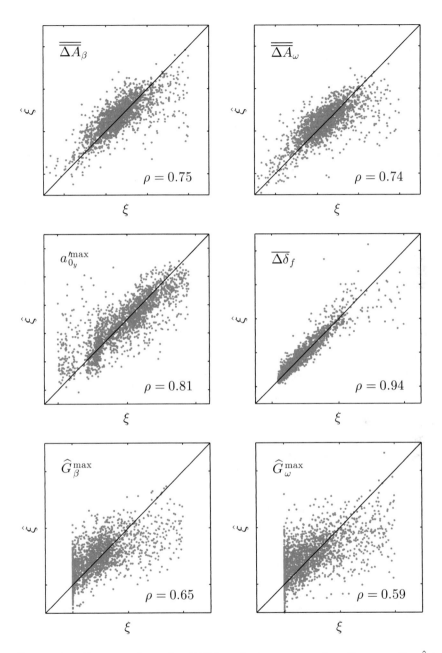

Figure 6.5: Scatter plots for RSMs relating approximation results $\hat{\xi}$ to original data ξ

resulting in 20000 design evaluations with RSM. Some of the designs resulting from this optimum search are then selected and concatenated to the already evaluated designs after an exact evaluation of criteria (5.11)–(5.14) based on the simulation model in Fig. 5.1. The RSMs are re-trained over the augmented set of evaluated designs resulting in a local refinement of the RSMs. The procedure stops if the number of refinement iterations N reaches a predefined value $N_{\mathrm{max}} = 5$ and terminates with the Pareto-sorting of all evaluated designs collected in the evaluated design set.

The selection procedure of update candidates has to fulfil both performance and space-filling criteria. The former implies the selection of the best performing designs, while the latter requires the selection of candidates located as far as possible from already evaluated designs. In order to find

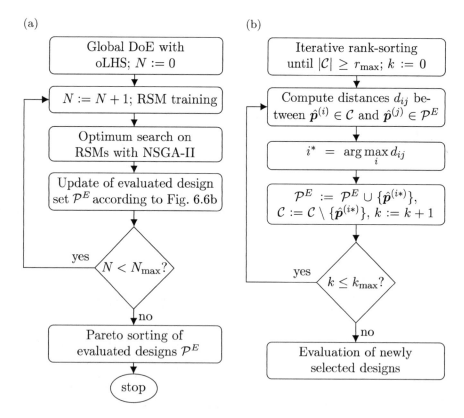

Figure 6.6: Optimisation procedure with (a) adaptive RSMs and (b) update of evaluated design set

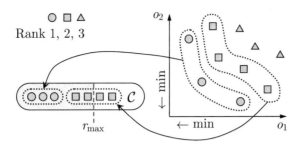

Figure 6.7: Illustration of the rank-sorting procedure in $\mathbb{R}^2$ criterion space

a compromise, the procedure in Fig. 6.6b is applied. Firstly, the 20000 designs generated by the optimum search in Fig. 6.6a are rank-sorted iteratively, i.e., the dominant designs are collected into the candidate set $\mathcal{C}$ until the number of rank-sorted designs reaches the arbitrarily chosen limit $r_{\max} = 300$, see Fig. 6.7. This provides a large enough pool of best performance candidates for further selection. Secondly, we may compute the Euclidean distances

$$d_{ij} = \sqrt{\sum_{k=1}^{16} \left(\frac{\hat{p}_k^{(i)} - \hat{p}_k^{(j)}}{\hat{p}_k^u - \hat{p}_k^l} \right)^2} \tag{6.16}$$

of all candidate points $\hat{\boldsymbol{p}}^{(i)} \in \mathcal{C}$ to all designs $\hat{\boldsymbol{p}}^{(j)} \in \mathcal{P}^E$ of already evaluated designs in the normalised design space. The most distant design $\hat{\boldsymbol{p}}^{(i*)}$ is removed from candidate set $\mathcal{C}$ and added to set $\mathcal{P}^E$. Subsequently, the remaining candidates are assessed and treated in the same way, where redundant distance computations are avoided, until the number k of selected designs reaches at least $k_{\max} = 100$. All these selected candidate points are evaluated by exact simulation which concludes the update process.

The computational time of the whole optimisation procedure with $N_{\max} = 5$ RSM-updates is

$$T_{\mathrm{RSM}}^{\times} + N_{\max}(T_{\mathrm{RSM}} + T_{\mathrm{NSGA-II}} + T_{\mathrm{eval}}) \approx 15620\,\mathrm{s} \approx 4.3\,\mathrm{h}, \tag{6.17}$$

where $T_{\mathrm{RSM}}^{\times} \approx 1610\,\mathrm{s}$ is the duration of RSM cross-validation only performed for auxiliary information, $T_{\mathrm{RSM}} \approx 82\,\mathrm{s}$ is the RSM training runtime, $T_{\mathrm{NSGA-II}} \approx 100 \cdot 200 \cdot 0.086\,\mathrm{s} = 1720\,\mathrm{s}$ summarises the duration of NSGA-II search on the RSMs following 200 individuals in the population of 100 generations, resulting from $0.086\,\mathrm{s}$ for each single RSM-evaluation, and $T_{\mathrm{eval}} \approx k_{\max} \cdot 10\,\mathrm{s} = 1000\,\mathrm{s}$ is the simulation run-time of the $k_{\max} = 100$

selected candidates where simulation of a single candidate takes about 10 s. The same search procedure with direct design evaluations of the 200 individuals in each generation would take approximately $500 \cdot 200 \cdot 10\,\text{s} = 10^6\,\text{s}$, assuming that NSGA-II has to run over $100 \cdot N_{\text{max}} = 500$ generations to obtain the same result quality. Thus the application of the RSMs speeds up optimisation by a factor of about 64.

6.4 Discussion of Optimisation Results

The main result of the optimisation procedure may be seen in Fig. 6.8 showing the finally resulting 44 Pareto-optimal elements of the set of evaluated designs in comparison to the manual design $\mathcal{M}$ corresponding to the parameter values given in Tables 5.1 and 5.2. We may conclude that many of the Pareto-optimal designs resulting from the above optimisation procedure dominate $\mathcal{M}$ in the sense that at least one objective value is lower without worsening the others.

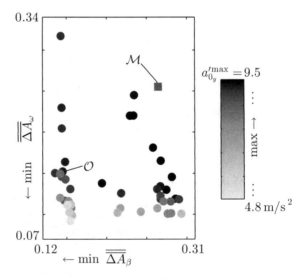

Figure 6.8: Pareto-optimal designs compared to a manual design $\mathcal{M}$ and selected Pareto-optimal design $\mathcal{O}$

For further investigations, a particular Pareto-optimal design $\mathcal{O}$ is selected from Fig. 6.8 and compared to $\mathcal{M}$ in Table 6.3. Obviously $\mathcal{O}$ has better values than $\mathcal{M}$ for all three criteria and fulfils all three constraints. This can also be seen by comparing the frequency responses of $\mathcal{M}$ and $\mathcal{O}$ in Fig. 6.9 where the optimised design $\mathcal{O}$ provides a generally better tracking performance as its cutoff frequencies for both transfer functions are higher than those of $\mathcal{M}$ bringing it closer to the desired characteristics visualised by dashed lines. This may also be observed in improved step responses in Fig. 6.10a, where design $\mathcal{O}$ is clearly faster than $\mathcal{M}$.

Some of the optimised parameters may explain the improved tracking performance. For example, state weights $\hat{q}_3$ and $\hat{q}_4$ in Table 6.3a are significantly higher for $\mathcal{O}$ than for $\mathcal{M}$. Since the ratio of the weights in Eq. (4.6) determines the controller behaviour, the higher values of $\hat{q}_3$ and $\hat{q}_4$ associated with β and ω make the controller faster to reduce deviations in these quantities by applying higher traction forces resulting in a faster tracking behaviour. Further, t_f in Table 6.3b is positive for optimised design $\mathcal{O}$ which results in an unstable linkage dynamics in contrast to $\mathcal{M}$. Unstable linkage dynamics eventually results in reduction of control effort and thus in faster response, as the lateral tyre forces tend to off-steer the wheels rather than centering them. The stability of the vehicle is ensured by the controller, though.

Table 6.3: Control (a) and mechanical (b) parameters as well as criterion values (c) of designs $\mathcal{M}$ and $\mathcal{O}$ (values given in SI-units)

(a)

	$p_c^{(\mathcal{M})}$	$p_c^{(\mathcal{O})}$
$\hat{q}_1$	-3	2.628
$\hat{q}_2$	-3	2.500
$\hat{q}_3$	6.699	9.873
$\hat{q}_4$	5	7.669
$\hat{q}_5$	-3	0.058
$\hat{q}_6$	-3	-0.253
$\hat{r}_1$	-3	-2.558
$\hat{r}_3$	-3	-2.699

(b)

	$p_m^{(\mathcal{M})}$	$p_m^{(\mathcal{O})}$
t_f	-0.010	0.018
t_r	0.045	0.076
r_f	0.078	0.095
r_r	-0.075	-0.074
c_f	1	69
c_r	2000	5512
d	80	69
a	0.703	0.700
c	0.088	0.006
h_f	0.140	0.183
h_r	0.140	0.150

(c)

	$\mathcal{M}$	$\mathcal{O}$
$\overline{\Delta A_\beta}$	0.264	0.145
$\overline{\Delta A_\omega}$	0.255	0.152
$a_{0_y}'^{\,\mathrm{max}}$	8.776	9.343
$\Delta \delta_f$	0.020	0.025
G_β^{max}	1.000	1.000
G_ω^{max}	1.000	1.000

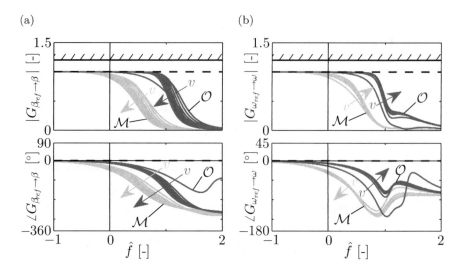

Figure 6.9: Comparison of manual design $\mathcal{M}$ (light) with optimised design $\mathcal{O}$ (dark) w.r.t. frequency responses of transfer functions (a) $G_{\beta_{\mathrm{ref}}\to\beta}$ and (b) $G_{\omega_{\mathrm{ref}}\to\omega}$ for vehicle speeds $v = 10\ldots 100\,\mathrm{km/h}$

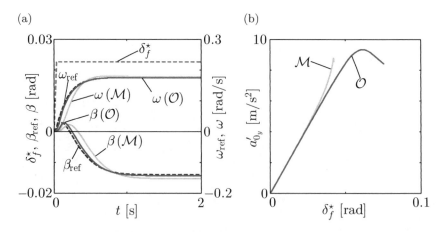

Figure 6.10: Step responses (a) for $v = 80\,\mathrm{km/h}$ (solid lines) compared to the references (black dashed) and steady-state cornering performance (b) of manual ($\mathcal{M}$, light), and optimised ($\mathcal{O}$, dark) designs

Also with respect to $a_{0_y}'^{\max}$, design $\mathcal{O}$ is superior compared to design $\mathcal{M}$ as a_{0_y}' reaches a significantly higher value in Fig. 6.10b. A further advantage is that design $\mathcal{O}$ provides an understeering behaviour as it encounters its limit, in contrary to design $\mathcal{M}$ which has a slight oversteering tendency as may be observed from its progressive $a_{0_y}'(\delta_f^\star)$ characteristic. Regarding the mean tracking error for low-speed manoeuvring in Fig. 6.11, design $\mathcal{O}$ is slightly inferior compared to design $\mathcal{M}$, although the difference is barely noticeable in the figure.

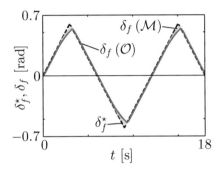

Figure 6.11: Low-speed performance of manual design $\mathcal{M}$ (light) compared to optimised design $\mathcal{O}$ (dark)

One may notice that the rear scrub radius r_r is negative for both designs $\mathcal{M}$ and $\mathcal{O}$ implying that the kingpin axis is located outside the wheels' centre line. Such suspension kinematics may be realised by a virtual kingpin axis resulting from, e.g., a four-bar linkage suspension mechanism shown in Fig. 6.12.

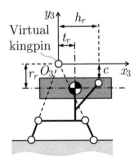

Figure 6.12: Suspension arrangement with virtual kingpin axis

7 Disturbance Rejection of the Differential Steering System

One may be concerned how the differential steering system without any dedicated steering actuator reacts to external disturbances. There are various types of external disturbance forces which may induce unintended steering response of the vehicle, such as lateral forces resulted from crosswind or road banking, or dynamic collision forces resulted from driving over road surface defects and obstacles, like potholes and curbs, for example. From the differential steering point of view, the latter disturbances have a greater importance as those act directly on the steering linkage and thus may result in more excessive unintended steering of the vehicle than, e.g., crosswind.

Although the disturbance rejection properties of differential steering systems is a niche research topic, there is some relevant literature available already. OKE AND NGUANG (2020) propose a robust control method based on H_∞ and fuzzy output feedback for crosswind disturbance rejection of a skid-steered vehicle. CAO ET AL. (2020) present a post-impact control algorithm for a vehicle with differential steering and impact disturbance, where the controller tries to keep the vehicle's stability and avoid secondary collisions by predictive path planning and actuating the differential steering system. WANG ET AL. (2020a) introduce active disturbance rejection control for DDAS where all disturbances are cumulated into an additional state variable, while WANG ET AL. (2020b) propose a H_∞-based robust control for DDAS where the external disturbance results from the road roughness and also parameter uncertainties of the suspension and sensor noise are taken into account.

This chapter investigates the situation where the vehicle drives over a curb with 50% overlap, i.e., only with its left wheels 1 and 3, see Fig. 7.1a. This produces dynamic collision forces $\boldsymbol{F}^{\mathrm{d}}$ during the wheel-curb collisions according to Fig. 7.1b. Since only the left wheels collide with the curb, the longitudinal component F_x^{d} of the collision forces produces asymmetric longitudinal forces on the wheels, resulting in steering moments about the kingpins according to the fundamental principle of differential steering. After introducing a wheel-curb collision model, the simulation framework is set up for collision investigations, and finally collision simulations are discussed.

(a)   (b)

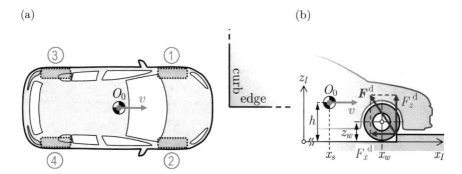

Figure 7.1: Vehicle rolling over a curb with its left wheels: (a) top view and (b) dynamic collision forces during wheel-curb collision of left wheels

7.1 Wheel-Curb Collision Model

In order to investigate the wheel-curb collision, we have to construct a proper dynamic model. In the literature, wheel-curb collision is usually investigated based on complex finite element models, see CHO ET AL. (2005) or MOUSSEAU AND HULBERT (1996). Since here the collision is only part of a vehicle response simulation, a simplified wheel-curb collision model is satisfactory.

According to Fig. 7.2, motion of the colliding wheel is described by the coordinates x_w, z_w and φ_w with respect to the inertial frame $I\{O_I, x_I, y_I, z_I\}$. The wheel contacts the curb edge first at position $x_{w_0} = x_e - \sqrt{2r_D h_e - h_e^2}$, $z_{w_0} = r_D$ and $\varphi_{w_0} = 0$, where x_e and h_e are the longitudinal distance from O_I and the height of the curb edge, respectively, and r_D is the dynamic rolling radius of the wheel. The velocity of the wheel is described by the velocity vector $\boldsymbol{v}_w$ consisting of longitudinal and vertical components v_{w_x} and v_{w_z}, respectively. The wheel rotates with angular velocity Ω. Since no suspension model is utilised, the vertical load forces $F_z = F_{z_f}/2$

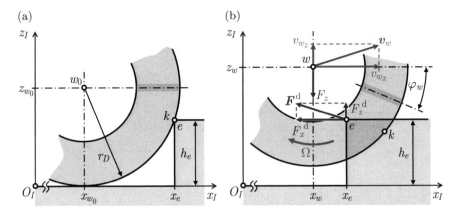

Figure 7.2: Wheel-curb collision model (a) in its initial configuration and (b) during collision

or $F_z = F_{z_r}/2$ are applied here as suspension forces, see also Fig. 2.7. The collision (disturbance) force $\boldsymbol{F}^{\mathrm{d}}$ is determined from the overlap of a hypothetical contact point k of the undeformed wheel and the curb, and from the radial stiffness c_w of the tyre. Traction force F_{x_1} or F_{x_3} is assumed to be zero at the instant of collision hence not included in the model. Influence of traction forces after the collision is discussed later in Section 7.2. The mass and inertia of the wheel is characterised by m_w and J_{w_r}, respectively, see also Section 2.2 and Table 5.1c.

In order to derive the EoM of the above model, let us define the position vector of the wheel as

$$\boldsymbol{r}_w = \begin{bmatrix} x_w & 0 & z_w \end{bmatrix}^{\mathrm{T}} \tag{7.1}$$

resulting in velocity

$$\boldsymbol{v}_w = \begin{bmatrix} v_{w_x} & 0 & v_{w_z} \end{bmatrix}^{\mathrm{T}} = \dot{\boldsymbol{r}}_w = \begin{bmatrix} \dot{x}_w & 0 & \dot{z}_w \end{bmatrix}^{\mathrm{T}} \tag{7.2}$$

and acceleration

$$\boldsymbol{a}_w = \dot{\boldsymbol{v}}_w = \begin{bmatrix} \dot{v}_{w_x} & 0 & \dot{v}_{w_z} \end{bmatrix}^{\mathrm{T}}. \tag{7.3}$$

For the sake of simplicity, let us assume that the longitudinal velocity v_{w_x} is determined by the vehicle speed v as the wheels are connected to the vehicle body. Assuming $v = \mathrm{const.}$ as defined in Section 2.1 despite the impact, this implies the constraints

$$\begin{aligned} v_{w_x} - v &= 0, \\ \dot{v}_{w_x} &= 0. \end{aligned} \tag{7.4}$$

Besides the translational motion, rolling of the wheel is characterised by the rotation angle φ_w resulting in the rotation matrix

$$S_w = \begin{bmatrix} \cos\varphi_w & 0 & \sin\varphi_w \\ 0 & 1 & 0 \\ -\sin\varphi_w & 0 & \cos\varphi_w \end{bmatrix}. \tag{7.5}$$

The angular velocity $\Omega = \dot{\varphi}_w$ and acceleration $\dot{\Omega}$ of the rolling wheel follow from Fig. 7.2 as

$$\boldsymbol{\omega}_w = \begin{bmatrix} 0 & \Omega & 0 \end{bmatrix}^{\mathrm{T}} = \begin{bmatrix} 0 & \dot{\varphi}_w & 0 \end{bmatrix}^{\mathrm{T}}, \tag{7.6}$$

$$\boldsymbol{\alpha}_w = \dot{\boldsymbol{\omega}}_w = \begin{bmatrix} 0 & \dot{\Omega} & 0 \end{bmatrix}^{\mathrm{T}}. \tag{7.7}$$

For the calculation of the collision force, we first have to express the position of the contact point k relative to the curb edge as

$$r_{ek} = r_w + S_w r'_{e_0} - r_e, \tag{7.8}$$

where $r'_{e_0} = \begin{bmatrix} x_e - x_{w_0} & 0 & h_e - z_{w_0} \end{bmatrix}^{\mathrm{T}}$ and $r_e = \begin{bmatrix} x_e & 0 & h_e \end{bmatrix}^{\mathrm{T}}$. Secondly, we have to consider the contact conditions of the wheel. It is assumed that the wheel is in contact with the curb edge if its distance from e is smaller than its radius and – for avoiding positive longitudinal collision force – the longitudinal component of the contact position vector r_{ek} is positive. From these considerations and Eq. (7.8), the collision force follows as

$$\boldsymbol{F}^{\mathrm{d}} = \begin{bmatrix} F_x^{\mathrm{d}} & 0 & F_z^{\mathrm{d}} \end{bmatrix}^{\mathrm{T}} = \begin{cases} -r_{ek} c_w & \text{for } \|r_{we}\| \le r_D \wedge r_{ek}^{(1)} \ge 0, \\ 0 & \text{otherwise,} \end{cases} \tag{7.9}$$

where $r_{we} = r_e - r_w = \begin{bmatrix} x_e - x_w & 0 & h_e - z_w \end{bmatrix}^{\mathrm{T}}$ is the vector from the wheel centre to the curb edge. As the line of action of the collision force $\boldsymbol{F}^{\mathrm{d}}$ does not cross the wheel's CoG in general, it produces the torque

$$M^{\mathrm{d}} = \boldsymbol{e}_y^{\mathrm{T}} \left(r_{we} \times \boldsymbol{F}^{\mathrm{d}} \right). \tag{7.10}$$

By combining the velocity definitions (7.2) and (7.6), the accelerations (7.3) and (7.7), the contraints (7.4), the vertical component F_z^{d} of the collision force (7.9) and the torque (7.10), we may deduce the EoM of the wheel as

$$\dot{x}_w = v,$$
$$\dot{z}_w = v_{w_z},$$
$$\dot{\varphi}_w = \Omega, \tag{7.11}$$
$$m_w \dot{v}_{w_z} = F_z^{\mathrm{d}} - F_z,$$
$$J_{w_r} \dot{\Omega} = M^{\mathrm{d}}.$$

In order to investigate the behaviour of the model, simulations are performed for the following conditions. It is assumed that the vertical load of the wheel is constant, i.e., $F_z = F_{z_f}/2 = 4155\,\mathrm{N}$, see also Eq. (2.68). The radial tyre stiffness is $c_w = 300\,\mathrm{kN/m}$. The EoM (7.11) is solved with the initial conditions $x_w(0) = x_{w_0}$, $z_w(0) = z_{w_0}$, $\varphi_w(0) = 0$, $v_{w_z}(0) = 0$ and $\Omega(0) = v/r_D$. The simulation results can be seen in Figs. 7.3, 7.4 for various velocities and curb heights. For the sake of simplicity, only the longitudinal component F_x^{d} of the collision force is investigated as that characterises the disturbance posed to the steering system by the collision. We may conclude that the longitudinal collision force depends strongly on the curb height, while the velocity has only a weaker influence, although

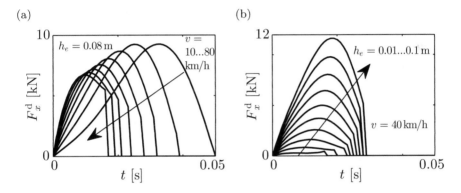

(a)   (b)

Figure 7.3: Longitudinal collision force for varying (a) speeds v and (b) curb height h_e

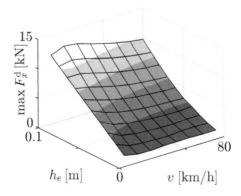

Figure 7.4: Peak values of the longitudinal collision force

still clearly noticeable. We may also investigate the motion of the wheel in a particular simulation case where it suddenly starts moving upwards until it leaves the contact, i.e., jumps over the curb, see Fig. 7.5.

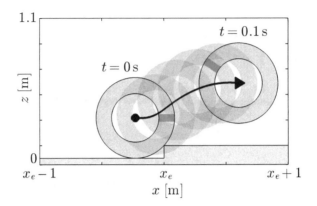

Figure 7.5: Typical wheel motion during the wheel-curb collision for $v = 30\,\mathrm{km/h}$ and $h_e = 0.1\,\mathrm{m}$

7.2 Simulation Framework for Collision Investigations

Besides the standalone investigation of the wheel-curb collision model, let us integrate it with the vehicle model for the analysis of the disturbance rejection behaviour of the differential steering system by joint simulation. The discussed collision model is applied for both front and rear wheels, where we have to establish the connection between the vehicle model and the wheel-curb collision model by replacing the velocity constraint (7.4) with

$$v_{w_x} - v_i^{(1)} = 0, \tag{7.12}$$

and substituting $F_z = F_{z_i}$ in the EoM (7.11), where $i \in \{1, 3\}$ as only the left side wheels collide with the curb; $v_i^{(1)}$ is the longitudinal component of individual wheel velocities (2.13), $F_{z_1} = F_{z_f}/2 = 4155\,\mathrm{N}$, and $F_{z_3} = F_{z_r}/2 = 2761\,\mathrm{N}$. In addition, the longitudinal collision forces of the wheels should be applied to the vehicle model. For this purpose, let us define the vector

$$\widehat{\boldsymbol{F}}_x^{\mathrm{d}} = \begin{bmatrix} F_{x_1}^{\mathrm{d}} & \cdots & F_{x_4}^{\mathrm{d}} \end{bmatrix}^{\mathrm{T}} \tag{7.13}$$

for wheel-individual longitudinal collision forces assuming that $F_{x_2}^{\mathrm{d}} = F_{x_4}^{\mathrm{d}} = 0 = \mathrm{const}$. Further, we have to consider the traction forces. In order to avoid confusion, let us distinguish between the actuating forces $\hat{\boldsymbol{u}} = \begin{bmatrix} F_{x_1} \dots F_{x_4} \end{bmatrix}^{\mathrm{T}}$ produced by the in-wheel motors and plant inputs $\boldsymbol{u}$. For the sake of simplicity, it is assumed that forces $\hat{\boldsymbol{u}}$ act entirely on the wheels and are not influenced by the collisions. Additionally, we have to check whether the traction forces can be applied by introducing a second contact condition, i.e., if the wheels jump off losing contact with the ground:

$$\hat{k}_{w_i} = \begin{cases} 1 & \text{for } z_{w_i} \leq r_D,\ i \in \{1,3\}, \\ 0 & \text{otherwise}, \end{cases} \tag{7.14}$$

$$\hat{k}_{w_2} = \hat{k}_{w_4} = 1 = \mathrm{const}.$$

By summarising the contact conditions (7.14) as $\hat{\boldsymbol{k}}_w = \begin{bmatrix} \hat{k}_{w_1} \dots \hat{k}_{w_4} \end{bmatrix}^{\mathrm{T}}$, the actual input forces follow as

$$\boldsymbol{u} = \hat{\boldsymbol{k}}_w \odot \hat{\boldsymbol{u}} - \widehat{\boldsymbol{F}}_x^{\mathrm{d}}, \tag{7.15}$$

where $\odot$ denotes the Hadamard product.

The block diagram of the joint simulation model is shown in Fig. 7.6. It is important to note, that the linearised vehicle model (3.32) is applied here instead of the nonlinear EoM (2.36). The reason for this choice is that the nonlinear tyre model may encounter numerical problems when excessive longitudinal forces are applied – see, e.g., Eq. (2.47) – and this way the longitudinal collision forces can be applied as input disturbances by utilising that the principle of superposition is applicable to linear systems.

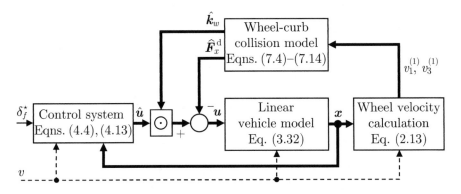

Figure 7.6: Vehcile model extended with wheel-curb collision model

7.3 Collision Simulations

Joint simulations are performed with parameter sets $\mathcal{M}$ and $\mathcal{O}$ introduced in Table 6.3, and their disturbance rejection behaviours are compared. For the simulations, a particular vehicle velocity $v = 40\,\mathrm{km/h}$ and curb height $h_w = 0.08\,\mathrm{m}$ are chosen. The simulations are performed such that the driver tries to keep the vehicle on a straight trajectory by keeping $\delta_f^\star = 0 = \mathrm{const.}$ while the collision happens.

Figure 7.7 shows the longitudinal collision forces for the left front and rear wheels. The collision forces are only indistinguishably different for $\mathcal{M}$ and $\mathcal{O}$ which is not the case for the traction forces of the front and rear wheels in Fig. 7.8. Note that the left forces F_{x_1} and F_{x_3} are zero in certain periods as the wheels jump off, i.e., due to multiplication by factors (7.14). A further outcome is that traction forces of optimised design $\mathcal{O}$ are much higher than those of manual design $\mathcal{M}$, indicating a more aggressive control intervention as a result of the higher control weights, see Table 6.3a.

We may compare the state variables as well. Magnitudes of steering angles in Fig. 7.9 are lower for $\mathcal{O}$ than for $\mathcal{M}$ which may be concluded even more for sideslip angle and yaw rate in Fig. 7.10. The most representative result, however, is the lateral deviation of the vehicle's trajectory from the desired straight run in Fig. 7.11. In the case of $\mathcal{M}$, the rate of change of $y_s(t)$ is approximately $0.15\,\mathrm{m/s}$ which would slowly drift the vehicle out of its lane. For $\mathcal{O}$, the deviation with only $\dot{y}_s = 0.03\,\mathrm{m/s}$ is significantly lower.

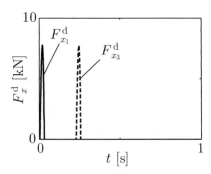

Figure 7.7: Longitudinal collision forces during joint simulation

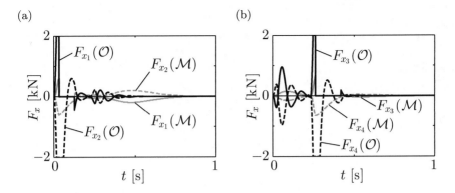

Figure 7.8: Traction forces applied to (a) front and (b) rear wheels

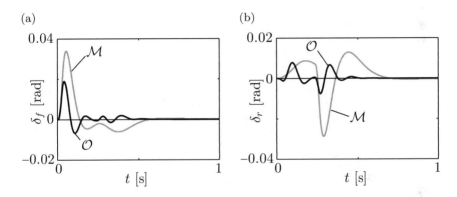

Figure 7.9: Front (a) and rear (b) steering angles during collision

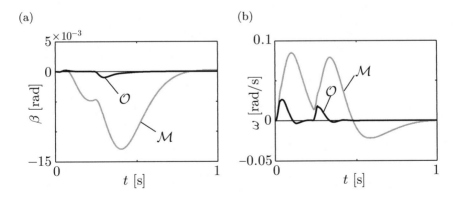

Figure 7.10: Sideslip angle (a) and yaw rate (b) during collision

Finally, we may conclude that designs $\mathcal{M}$ and $\mathcal{O}$ are both robust against disturbance forces resulting from wheel-curb collision. Although the vehicle deviates from its straight run path, the deviations are limited and might be easily countered by the driver, especially in the case of optimised design $\mathcal{O}$. We may also conclude that the aggressive control intervention contributes to disturbance rejection.

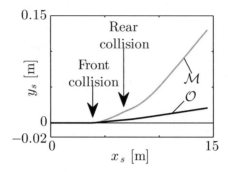

Figure 7.11: Vehicle trajectory $y_s(x_s)$ deviation

8 Conclusions and Outlook

The steering concept presented in this thesis challenges conventional steering systems by applying only passive steering linkages and traction force differences produced by in-wheel motors. The new steering concept provides a useful alternative to conventional steering and may be utilised in future vehicles equipped with four in-wheel electric motors. Advantages are cost saving by the lack of the additional conventional steering device, and more flexibility in the design. Omission of dedicated steering devices simplifies the chassis design and saves design space, which allows placing of axles and wheels with more design freedom.

A planar multi-body model taking into account the dynamic interaction between the steering linkages and the vehicle body, and nonlinear tyre models with an emphasis on standstill behaviour are created for studying the differential steering principle. Associated control systems for steering control via longitudinal traction forces are developed for both low- and high-speed regimes. For higher speeds, a linear full state feedback control system is applied, whereas a simple angle tracking controller is used for operation at lower speeds. Various simulation experiments with a manually selected parameter set demonstrate that the steering performance of the proposed differential steering concept is comparable to that of conventional passenger cars.

In order to design control concepts, it is necessary first to develop a symbolic linearisation method for complex multi-body models with a combination of non-holonomic constraints and kinematic loops. One of the specific problems to be resolved is avoidance of differentiation of inverted matrices as part of the linearisation, in order to obtain the linearised equations in a computationally feasible and efficient way. Simulations of both linear and nonlinear vehicle models show that the proposed method is applicable for the problem, and the results of both models are comparable with a reasonable accuracy.

Since experience is missing for designing such a new concept of differential steering, a multi-objective optimisation procedure is developed to prove the general applicability of the proposed concept and to find a setup where it is able to perform as well as, or even outperform conventional vehicles with dedicated steering devices. It is revealed that optimised designs feature high state weights of their control system in combination with a mechanically unstable linkage configuration of their front axle to achieve an overall stable behaviour with fast dynamics comparable with those of classical steering solutions. For optimisation, three design objectives and three constraints are deduced from standard vehicle dynamics test manoeuvres to assess low-speed manoeuvring, quasi-static cornering and high-speed dynamic steering performance. A multi-objective optimisation algorithm is combined with neural network-based response surface models to find multiple Pareto-optimal design trade-offs in a reasonable time which mostly dominate the intuitively found design.

Finally, it is proven that the proposed differential steering concept is immune to external disturbances. For this purpose, a simulation model for one-sided wheel-curb collision is developed, since this may be regarded as the most critical manoeuvre where the collision force directly acts on the steering linkage with maximum steering moment and thus has a significant impact on steering behaviour. However, the results show that the proposed differential steering concept and the associated control algorithm effectively reject such disturbances resulting in a safe operation of the vehicle.

Since the differential steering principle is still an emerging idea, there are numerous opportunities for further improvements. For example, the vehicle's behaviour could be investigated more precisely by using spatial multi-body models and taking into account additional details regarding a dynamic tyre model. The design procedure may take into account parameter uncertainties in order to generate robust designs. Regarding the control system, ideal measurement of state variables is assumed and the application of state estimators is not discussed in this thesis, although those may have an essential effect on vehicle dynamics control. Thus, application of various state estimation techniques could be a natural extension of the present work. Further, more advanced control methods like feedback linearisation or μ-synthesis could be applied in future investigations, providing improved performance and robustness. Last but not least, experimental validation of the proposed steering concept should augment our knowledge, where the results of the thesis may act as a good starting point for building a prototype vehicle.

Appendix: Detailed Results of Model Derivation

The Appendix presents the details of the multi-body vehicle model (2.36) by showing the results of the main derivation steps from Sections 2.1 and 2.2. Derivation results of the kinematics are described first, followed by the details of the EoM and the constraints.

A.1 Kinematics

Position vectors of bodies from Eq. (2.8) read as

$$\boldsymbol{r}_0 = \begin{bmatrix} x_s & y_s & 0 \end{bmatrix}^{\mathrm{T}}, \tag{A.1}$$

$$\boldsymbol{r}_1 = \begin{bmatrix} x_s + t_f \cos(\psi + \delta_1) - r_f \sin(\psi + \delta_1) + l_f \cos\psi - a\sin\psi \\ y_s + r_f \cos(\psi + \delta_1) + t_f \sin(\psi + \delta_1) + a\cos\psi + l_f \sin\psi \\ 0 \end{bmatrix}, \tag{A.2}$$

$$\boldsymbol{r}_2 = \begin{bmatrix} x_s + t_f \cos(\psi + \delta_2) + r_f \sin(\psi + \delta_2) + l_f \cos\psi + a\sin\psi \\ y_s - r_f \cos(\psi + \delta_2) + t_f \sin(\psi + \delta_2) - a\cos\psi + l_f \sin\psi \\ 0 \end{bmatrix}, \tag{A.3}$$

$$\boldsymbol{r}_3 = \begin{bmatrix} x_s + t_r \cos(\psi + \delta_3) - r_r \sin(\psi + \delta_3) - l_r \cos\psi - a\sin\psi \\ y_s + r_r \cos(\psi + \delta_3) + t_r \sin(\psi + \delta_3) + a\cos\psi - l_r \sin\psi \\ 0 \end{bmatrix}, \tag{A.4}$$

$$\boldsymbol{r}_4 = \begin{bmatrix} x_s + t_r \cos(\psi + \delta_4) + r_r \sin(\psi + \delta_4) - l_r \cos\psi + a\sin\psi \\ y_s - r_r \cos(\psi + \delta_4) + t_r \sin(\psi + \delta_4) - a\cos\psi - l_r \sin\psi \\ 0 \end{bmatrix}. \tag{A.5}$$

Rotation matrices of the bodies result from Eq. (2.9) as

$$
\boldsymbol{S}_0 =
\begin{bmatrix}
\cos(\psi) & -\sin(\psi) & 0 \\
\sin(\psi) & \cos(\psi) & 0 \\
0 & 0 & 1
\end{bmatrix},
$$

$$
\boldsymbol{S}_i =
\begin{bmatrix}
\cos(\psi + \delta_i) & -\sin(\psi + \delta_i) & 0 \\
\sin(\psi + \delta_i) & \cos(\psi + \delta_i) & 0 \\
0 & 0 & 1
\end{bmatrix}, \quad i = 1 \ldots 4.
$$

(A.6)

Velocities of bodies' CoGs are obtained from Eqns. (2.12)–(2.14) as

$$
\boldsymbol{v}_0 = \begin{bmatrix} v\cos(\beta + \psi) & v\sin(\beta + \psi) & 0 \end{bmatrix}^{\mathrm{T}},
$$

(A.7)

$$
\boldsymbol{v}_1 =
\begin{bmatrix}
v\cos(\beta + \psi) - \omega(r_f \cos(\psi + \delta_1) + t_f \sin(\psi + \delta_1) + a\cos\psi + l_f \sin\psi) \\
\quad - \dot{\delta}_1(r_f \cos(\psi + \delta_1) + t_f \sin(\psi + \delta_1)) \\[2mm]
v\sin(\beta + \psi) + \omega(t_f \cos(\psi + \delta_1) - r_f \sin(\psi + \delta_1) + l_f \cos\psi - a\sin\psi) \\
\quad + \dot{\delta}_1(t_f \cos(\psi + \delta_1) - r_f \sin(\psi + \delta_1)) \\[2mm]
0
\end{bmatrix},
$$

(A.8)

$$
\boldsymbol{v}_2 =
\begin{bmatrix}
v\cos(\beta + \psi) + \omega(r_f \cos(\psi + \delta_2) - t_f \sin(\psi + \delta_2) + a\cos\psi - l_f \sin\psi) \\
\quad + \dot{\delta}_2(r_f \cos(\psi + \delta_2) - t_f \sin(\psi + \delta_2)) \\[2mm]
v\sin(\beta + \psi) + \omega(t_f \cos(\psi + \delta_2) + r_f \sin(\psi + \delta_2) + l_f \cos\psi + a\sin\psi) \\
\quad + \dot{\delta}_2(t_f \cos(\psi + \delta_2) + r_f \sin(\psi + \delta_2)) \\[2mm]
0
\end{bmatrix},
$$

(A.9)

$$
\boldsymbol{v}_3 =
\begin{bmatrix}
v\cos(\beta + \psi) - \omega(r_r \cos(\psi + \delta_3) + t_r \sin(\psi + \delta_3) + a\cos\psi - l_r \sin\psi) \\
\quad - \dot{\delta}_3(r_r \cos(\psi + \delta_3) + t_r \sin(\psi + \delta_3)) \\[2mm]
v\sin(\beta + \psi) + \omega(t_r \cos(\psi + \delta_3) - r_r \sin(\psi + \delta_3) - l_r \cos\psi - a\sin\psi) \\
\quad + \dot{\delta}_3(t_r \cos(\psi + \delta_3) - r_r \sin(\psi + \delta_3)) \\[2mm]
0
\end{bmatrix},
$$

(A.10)

$$\boldsymbol{v}_4 =$$

$$\begin{bmatrix} v\cos\left(\beta+\psi\right)+\omega(r_r\cos\left(\psi+\delta_4\right)-t_r\sin\left(\psi+\delta_4\right)+a\cos\psi+l_r\sin\psi) \\ +\dot{\delta}_4(r_r\cos\left(\psi+\delta_4\right)-t_r\sin\left(\psi+\delta_4\right)) \\[1em] v\sin\left(\beta+\psi\right)+\omega(t_r\cos\left(\psi+\delta_4\right)+r_r\sin\left(\psi+\delta_4\right)-l_r\cos\psi+a\sin\psi) \\ +\dot{\delta}_4(t_r\cos\left(\psi+\delta_4\right)+r_r\sin\left(\psi+\delta_4\right)) \\[1em] 0 \end{bmatrix}.$$

$$(A.11)$$

Body accelerations from Eqns. (2.15), (2.16) follow as

$$\boldsymbol{a}_0 = \begin{bmatrix} -\dot{\beta}v\sin\left(\beta+\psi\right) & \dot{\beta}v\cos\left(\beta+\psi\right) & 0 \end{bmatrix}^{\mathrm{T}} + \bar{\boldsymbol{a}}_0, \qquad (A.12)$$

$$\boldsymbol{a}_1 = \begin{bmatrix} -\ddot{\delta}_1(r_f\cos\left(\psi+\delta_1\right)+t_f\sin\left(\psi+\delta_1\right))-\dot{\omega}(r_f\cos\left(\psi+\delta_1\right) \\ +t_f\sin\left(\psi+\delta_1\right)+a\cos\psi+l_f\sin\psi)-\dot{\beta}v\sin\left(\beta+\psi\right) \\[1em] \ddot{\delta}_1(t_f\cos\left(\psi+\delta_1\right)-r_f\sin\left(\psi+\delta_1\right))+\dot{\omega}(t_f\cos\left(\psi+\delta_1\right) \\ -r_f\sin\left(\psi+\delta_1\right)+l_f\cos\psi-a\sin\psi)+\dot{\beta}v\cos\left(\beta+\psi\right) \\[1em] 0 \end{bmatrix} + \bar{\boldsymbol{a}}_1,$$

$$(A.13)$$

$$\boldsymbol{a}_2 = \begin{bmatrix} \ddot{\delta}_2(r_f\cos\left(\psi+\delta_2\right)-t_f\sin\left(\psi+\delta_2\right))+\dot{\omega}(r_f\cos\left(\psi+\delta_2\right) \\ -t_f\sin\left(\psi+\delta_2\right)+a\cos\psi-l_f\sin\psi)-\dot{\beta}v\sin\left(\beta+\psi\right) \\[1em] \ddot{\delta}_2(t_f\cos\left(\psi+\delta_2\right)+r_f\sin\left(\psi+\delta_2\right))+\dot{\omega}(t_f\cos\left(\psi+\delta_2\right) \\ +r_f\sin\left(\psi+\delta_2\right)+l_f\cos\psi+a\sin\psi)+\dot{\beta}v\cos\left(\beta+\psi\right) \\[1em] 0 \end{bmatrix} + \bar{\boldsymbol{a}}_2,$$

$$(A.14)$$

$$\boldsymbol{a}_3 = \begin{bmatrix} -\ddot{\delta}_3(r_r\cos\left(\psi+\delta_3\right)+t_r\sin\left(\psi+\delta_3\right))-\dot{\omega}(r_r\cos\left(\psi+\delta_3\right) \\ +t_r\sin\left(\psi+\delta_3\right)+a\cos\psi-l_r\sin\psi)-\dot{\beta}v\sin\left(\beta+\psi\right) \\[1em] \ddot{\delta}_3(t_r\cos\left(\psi+\delta_3\right)-r_r\sin\left(\psi+\delta_3\right))+\dot{\omega}(t_r\cos\left(\psi+\delta_3\right) \\ -r_r\sin\left(\psi+\delta_3\right)-l_r\cos\psi-a\sin\psi)+\dot{\beta}v\cos\left(\beta+\psi\right) \\[1em] 0 \end{bmatrix} + \bar{\boldsymbol{a}}_3,$$

$$(A.15)$$

$$\boldsymbol{a}_4 = \begin{bmatrix} \ddot{\delta}_4(r_r\cos{(\psi + \delta_4)} - t_r\sin{(\psi + \delta_4)}) + \dot{\omega}(r_r\cos{(\psi + \delta_4)} \\ - t_r\sin{(\psi + \delta_4)} + a\cos\psi + l_r\sin\psi) - \dot{\beta}v\sin{(\beta + \psi)} \\ \\ \ddot{\delta}_4(t_r\cos{(\psi + \delta_4)} + r_r\sin{(\psi + \delta_4)}) + \dot{\omega}(t_r\cos{(\psi + \delta_4)} \\ + r_r\sin{(\psi + \delta_4)} - l_r\cos\psi + a\sin\psi) + \dot{\beta}v\cos{(\beta + \psi)} \\ \\ 0 \end{bmatrix} + \bar{\boldsymbol{a}}_4$$

$$(A.16)$$

with local acceleration terms (2.16):

$$\bar{\boldsymbol{a}}_0 = \begin{bmatrix} -\omega v\sin{(\beta + \psi)} & \omega v\cos{(\beta + \psi)} & 0 \end{bmatrix}^{\mathrm{T}}, \qquad (A.17)$$

$$\bar{\boldsymbol{a}}_1 = \begin{bmatrix} -\omega v\sin{(\beta + \psi)} - (\dot{\delta}_1 + \omega)^2(t_f\cos{(\delta_1 + \psi)} - r_f\sin{(\delta_1 + \psi)}) \\ -\omega^2(l_f\cos\psi - a\sin\psi) \\ \\ \omega v\cos{(\beta + \psi)} - (\dot{\delta}_1 + \omega)^2(r_f\cos{(\delta_1 + \psi)} + t_f\sin{(\delta_1 + \psi)}) \\ -\omega^2(a\cos\psi + l_f\sin\psi) \\ \\ 0 \end{bmatrix},$$

$$(A.18)$$

$$\bar{\boldsymbol{a}}_2 = \begin{bmatrix} -\omega v\sin{(\beta + \psi)} - (\dot{\delta}_2 + \omega)^2(t_f\cos{(\delta_2 + \psi)} + r_f\sin{(\delta_2 + \psi)}) \\ -\omega^2(l_f\cos\psi + a\sin\psi) \\ \\ \omega v\cos{(\beta + \psi)} + (\dot{\delta}_2 + \omega)^2(r_f\cos{(\delta_2 + \psi)} - t_f\sin{(\delta_2 + \psi)}) \\ +\omega^2(a\cos\psi - l_f\sin\psi) \\ \\ 0 \end{bmatrix},$$

$$(A.19)$$

$$\bar{\boldsymbol{a}}_3 = \begin{bmatrix} -\omega v\sin{(\beta + \psi)} - (\dot{\delta}_3 + \omega)^2(t_r\cos{(\delta_3 + \psi)} - r_r\sin{(\delta_3 + \psi)}) \\ +\omega^2(l_r\cos\psi + a\sin\psi) \\ \\ \omega v\cos{(\beta + \psi)} - (\dot{\delta}_3 + \omega)^2(r_r\cos{(\delta_3 + \psi)} + t_r\sin{(\delta_3 + \psi)}) \\ -\omega^2(a\cos\psi - l_r\sin\psi) \\ \\ 0 \end{bmatrix},$$

$$(A.20)$$

$$\bar{\boldsymbol{a}}_4 = \begin{bmatrix} -\omega v\sin{(\beta + \psi)} - (\dot{\delta}_4 + \omega)^2(t_r\cos{(\delta_4 + \psi)} + r_r\sin{(\delta_4 + \psi)}) \\ +\omega^2(l_r\cos\psi - a\sin\psi) \\ \\ \omega v\cos{(\beta + \psi)} + (\dot{\delta}_4 + \omega)^2(r_r\cos{(\delta_4 + \psi)} - t_r\sin{(\delta_4 + \psi)}) \\ +\omega^2(a\cos\psi + l_r\sin\psi) \\ \\ 0 \end{bmatrix}.$$

$$(A.21)$$

Translational Jacobians in Eq. (2.16) read as

$$
\boldsymbol{L}_{T_0} = \begin{bmatrix} -v\sin\left(\beta+\psi\right) & 0 & 0 & 0 & 0 & 0 \\ v\cos\left(\beta+\psi\right) & 0 & 0 & 0 & 0 & 0 \\ 0 & 0 & 0 & 0 & 0 & 0 \end{bmatrix}, \tag{A.22}
$$

$$
\boldsymbol{L}_{T_1} = \begin{bmatrix} -v\sin\left(\beta+\psi\right) & \begin{array}{l} -r_f\cos\left(\psi+\delta_1\right) \\ -t_f\sin\left(\psi+\delta_1\right) \\ -a\cos\psi \\ -l_f\sin\psi \end{array} & \begin{array}{l} -r_f\cos\left(\psi+\delta_1\right) \\ -t_f\sin\left(\psi+\delta_1\right) \end{array} & 0 & 0 & 0 \\[4ex] v\cos\left(\beta+\psi\right) & \begin{array}{l} t_f\cos\left(\psi+\delta_1\right) \\ -r_f\sin\left(\psi+\delta_1\right) \\ +l_f\cos\psi \\ -a\sin\psi \end{array} & \begin{array}{l} t_f\cos\left(\psi+\delta_1\right) \\ -r_f\sin\left(\psi+\delta_1\right) \end{array} & 0 & 0 & 0 \\[4ex] 0 & 0 & 0 & 0 & 0 & 0 \end{bmatrix}, \tag{A.23}
$$

$$
\boldsymbol{L}_{T_2} = \begin{bmatrix} -v\sin\left(\beta+\psi\right) & \begin{array}{l} r_f\cos\left(\psi+\delta_2\right) \\ -t_f\sin\left(\psi+\delta_2\right) \\ +a\cos\psi \\ -l_f\sin\psi \end{array} & 0 & \begin{array}{l} r_f\cos\left(\psi+\delta_2\right) \\ -t_f\sin\left(\psi+\delta_2\right) \end{array} & 0 & 0 \\[4ex] v\cos\left(\beta+\psi\right) & \begin{array}{l} t_f\cos\left(\psi+\delta_2\right) \\ +r_f\sin\left(\psi+\delta_2\right) \\ +l_f\cos\psi \\ +a\sin\psi \end{array} & 0 & \begin{array}{l} t_f\cos\left(\psi+\delta_2\right) \\ +r_f\sin\left(\psi+\delta_2\right) \end{array} & 0 & 0 \\[4ex] 0 & 0 & 0 & 0 & 0 & 0 \end{bmatrix}, \tag{A.24}
$$

$$
\boldsymbol{L}_{T_3} = \begin{bmatrix} -v\sin\left(\beta+\psi\right) & \begin{array}{l} l_r\sin\psi \\ -t_r\sin\left(\psi+\delta_3\right) \\ -a\cos\psi \\ -r_r\cos\left(\psi+\delta_3\right) \end{array} & 0 & 0 & \begin{array}{l} -r_r\cos\left(\psi+\delta_3\right) \\ -t_r\sin\left(\psi+\delta_3\right) \end{array} & 0 \\[4ex] v\cos\left(\beta+\psi\right) & \begin{array}{l} t_r\cos\left(\psi+\delta_3\right) \\ -r_r\sin\left(\psi+\delta_3\right) \\ -l_r\cos\psi \\ -a\sin\psi \end{array} & 0 & 0 & \begin{array}{l} t_r\cos\left(\psi+\delta_3\right) \\ -r_r\sin\left(\psi+\delta_3\right) \end{array} & 0 \\[4ex] 0 & 0 & 0 & 0 & 0 & 0 \end{bmatrix}, \tag{A.25}
$$

$$
\boldsymbol{L}_{T_4} =
\begin{bmatrix}
-v\sin\left(\beta+\psi\right) &
\begin{matrix} r_r\cos\left(\psi+\delta_4\right) \\ -t_r\sin\left(\psi+\delta_4\right) \\ +a\cos\psi \\ +l_r\sin\psi \end{matrix}
& 0 & 0 & 0 &
\begin{matrix} r_r\cos\left(\psi+\delta_4\right) \\ -t_r\sin\left(\psi+\delta_4\right) \end{matrix} \\[2em]
v\cos\left(\beta+\psi\right) &
\begin{matrix} t_r\cos\left(\psi+\delta_4\right) \\ +r_r\sin\left(\psi+\delta_4\right) \\ -l_r\cos\psi \\ +a\sin\psi \end{matrix}
& 0 & 0 & 0 &
\begin{matrix} t_r\cos\left(\psi+\delta_4\right) \\ +r_r\sin\left(\psi+\delta_4\right) \end{matrix} \\[2em]
0 & 0 & 0 & 0 & 0 & 0
\end{bmatrix}.
$$

$$\text{(A.26)}$$

Angular velocities of the bodies result from Eq. (2.17) as

$$
\boldsymbol{\omega}_0 = \begin{bmatrix} 0 & 0 & \omega \end{bmatrix}^{\mathrm{T}}, \quad
\boldsymbol{\omega}_i = \begin{bmatrix} 0 & 0 & \omega + \dot{\delta}_i \end{bmatrix}^{\mathrm{T}}, \quad i = 1\ldots 4. \tag{A.27}
$$

Angular accelerations (2.18), (2.19) of the bodies are

$$
\boldsymbol{\alpha}_0 = \begin{bmatrix} 0 & 0 & \dot{\omega} \end{bmatrix}^{\mathrm{T}}, \quad
\boldsymbol{\alpha}_i = \begin{bmatrix} 0 & 0 & \dot{\omega} + \ddot{\delta}_i \end{bmatrix}^{\mathrm{T}}, \quad i = 1\ldots 4. \tag{A.28}
$$

Rotational Jacobians in Eq. (2.19) are

$$
\boldsymbol{L}_{R_0} = \begin{bmatrix} 0 & 0 & 0 & 0 & 0 & 0 \\ 0 & 0 & 0 & 0 & 0 & 0 \\ 0 & 1 & 0 & 0 & 0 & 0 \end{bmatrix}, \quad
\boldsymbol{L}_{R_1} = \begin{bmatrix} 0 & 0 & 0 & 0 & 0 & 0 \\ 0 & 0 & 0 & 0 & 0 & 0 \\ 0 & 1 & 1 & 0 & 0 & 0 \end{bmatrix},
$$

$$
\boldsymbol{L}_{R_2} = \begin{bmatrix} 0 & 0 & 0 & 0 & 0 & 0 \\ 0 & 0 & 0 & 0 & 0 & 0 \\ 0 & 1 & 0 & 1 & 0 & 0 \end{bmatrix}, \quad
\boldsymbol{L}_{R_3} = \begin{bmatrix} 0 & 0 & 0 & 0 & 0 & 0 \\ 0 & 0 & 0 & 0 & 0 & 0 \\ 0 & 1 & 0 & 0 & 1 & 0 \end{bmatrix}, \tag{A.29}
$$

$$
\boldsymbol{L}_{R_4} = \begin{bmatrix} 0 & 0 & 0 & 0 & 0 & 0 \\ 0 & 0 & 0 & 0 & 0 & 0 \\ 0 & 1 & 0 & 0 & 0 & 1 \end{bmatrix}.
$$

A.2 Equations of Motion

The EoM terms (2.27)–(2.29) are shown in decomposition as $\boldsymbol{M} = \sum_i \boldsymbol{M}_i$, $\boldsymbol{k} = \sum_i \boldsymbol{k}_i$ and $\boldsymbol{q} = \sum_i \boldsymbol{q}_i$ with components $\boldsymbol{M}_i$, $\boldsymbol{k}_i$ and $\boldsymbol{q}_i$, $i = 0\ldots 4$, resulting from the five bodies.

Generalised mass matrix components are $\boldsymbol{M}_i = \boldsymbol{L}_{T_i}^{\mathrm{T}} m_i \boldsymbol{L}_{T_i} + \boldsymbol{L}_{R_i}^{\mathrm{T}} \boldsymbol{I}_i \boldsymbol{L}_{R_i}$, where

$$\boldsymbol{M}_0 = \begin{bmatrix} m_b v^2 & 0 & 0 & 0 & 0 & 0 \\ 0 & J_{b_z} & 0 & 0 & 0 & 0 \\ 0 & 0 & 0 & 0 & 0 & 0 \\ 0 & 0 & 0 & 0 & 0 & 0 \\ 0 & 0 & 0 & 0 & 0 & 0 \\ 0 & 0 & 0 & 0 & 0 & 0 \end{bmatrix}, \tag{A.30}$$

$\boldsymbol{M}_1 =$

$$\begin{bmatrix} m_w v^2 & \begin{matrix} m_w v(l_f \cos\beta \\ + a\sin\beta \\ + t_f \cos(\beta - \delta_1) \\ + r_f \sin(\beta - \delta_1)) \end{matrix} & \begin{matrix} m_w v(t_f \cos(\beta - \delta_1) \\ + r_f \sin(\beta - \delta_1)) \end{matrix} & 0 & 0 & 0 \\[4ex] \begin{matrix} m_w v(l_f \cos\beta \\ + a\sin\beta \\ + t_f \cos(\beta - \delta_1) \\ + r_f \sin(\beta - \delta_1)) \end{matrix} & \begin{matrix} m_w(a^2 + l_f^2 \\ + r_f^2 + t_f^2 \\ + 2ar_f \cos\delta_1 \\ + 2l_f t_f \cos\delta_1 \\ + 2at_f \sin\delta_1 \\ - 2l_f r_f \sin\delta_1) \\ + J_w \end{matrix} & \begin{matrix} m_w(r_f^2 + t_f^2 \\ + ar_f \cos\delta_1 \\ + l_f t_f \cos\delta_1 \\ + at_f \sin\delta_1 \\ - l_f r_f \sin\delta_1) \\ + J_w \end{matrix} & 0 & 0 & 0 \\[4ex] \begin{matrix} m_w v(t_f \cos(\beta - \delta_1) \\ + r_f \sin(\beta - \delta_1)) \end{matrix} & \begin{matrix} m_w(r_f^2 + t_f^2 \\ + ar_f \cos\delta_1 \\ + l_f t_f \cos\delta_1 \\ + at_f \sin\delta_1 \\ - l_f r_f \sin\delta_1) \\ + J_w \end{matrix} & m_w(r_f^2 + t_f^2) + J_w & 0 & 0 & 0 \\[4ex] 0 & 0 & 0 & 0 & 0 & 0 \\ 0 & 0 & 0 & 0 & 0 & 0 \\ 0 & 0 & 0 & 0 & 0 & 0 \end{bmatrix}, \tag{A.31}$$

$$\boldsymbol{M_2} =$$

$$
\begin{bmatrix}
m_w v^2 & \begin{aligned} m_w v(l_f \cos\beta \\ - a\sin\beta \\ + t_f \cos(\beta-\delta_2) \\ - r_f \sin(\beta-\delta_2)) \end{aligned} & 0 & \begin{aligned} m_w v(t_f \cos(\beta-\delta_2) \\ - r_f \sin(\beta-\delta_2)) \end{aligned} & 0 & 0 \\[2ex]
\begin{aligned} m_w v(l_f \cos\beta \\ - a\sin\beta \\ + t_f \cos(\beta-\delta_2) \\ - r_f \sin(\beta-\delta_2)) \end{aligned} & \begin{aligned} m_w(a^2 + l_f^2 \\ + r_f^2 + t_f^2 \\ + 2ar_f \cos\delta_2 \\ + 2l_f t_f \cos\delta_2 \\ - 2at_f \sin\delta_2 \\ + 2l_f r_f \sin\delta_2) \\ + J_w \end{aligned} & 0 & \begin{aligned} m_w(r_f^2 + t_f^2 \\ + ar_f \cos\delta_2 \\ + l_f t_f \cos\delta_2 \\ - at_f \sin\delta_2 \\ + l_f r_f \sin\delta_2) \\ + J_w \end{aligned} & 0 & 0 \\[2ex]
0 & 0 & 0 & 0 & 0 & 0 \\[1ex]
\begin{aligned} m_w v(t_f \cos(\beta-\delta_2) \\ - r_f \sin(\beta-\delta_2)) \end{aligned} & \begin{aligned} m_w(r_f^2 + t_f^2 \\ + ar_f \cos\delta_2 \\ + l_f t_f \cos\delta_2 \\ - at_f \sin\delta_2 \\ + l_f r_f \sin\delta_2) \\ + J_w \end{aligned} & 0 & m_w(r_f^2 + t_f^2) + J_w & 0 & 0 \\[2ex]
0 & 0 & 0 & 0 & 0 & 0 \\[1ex]
0 & 0 & 0 & 0 & 0 & 0
\end{bmatrix},
$$

$$(A.32)$$

$$M_3 =$$

$$
\begin{bmatrix}
m_w v^2 & \begin{aligned} &- m_w v(l_r \cos\beta \\ &- a \sin\beta \\ &- t_r \cos(\beta - \delta_3) \\ &- r_r \sin(\beta - \delta_3)) \end{aligned} & 0 & 0 & \begin{aligned} m_w v(t_r \cos(\beta - \delta_3) \\ + r_r \sin(\beta - \delta_3)) \end{aligned} & 0 \\[3em]
\begin{aligned} &- m_w v(l_r \cos\beta \\ &- a \sin\beta \\ &- t_r \cos(\beta - \delta_3) \\ &- r_r \sin(\beta - \delta_3)) \end{aligned} & \begin{aligned} &m_w(a^2 + l_r^2 \\ &+ r_r^2 + t_r^2 \\ &+ 2ar_r \cos\delta_3 \\ &- 2l_r t_r \cos\delta_3 \\ &+ 2at_r \sin\delta_3 \\ &+ 2l_r r_r \sin\delta_3) \\ &+ J_w \end{aligned} & 0 & 0 & \begin{aligned} &m_w(r_r^2 + t_r^2 \\ &+ ar_r \cos\delta_3 \\ &- l_r t_r \cos\delta_3 \\ &+ at_r \sin\delta_3 \\ &+ l_r r_r \sin\delta_3) \\ &+ J_w \end{aligned} & 0 \\[3em]
0 & 0 & 0 & 0 & 0 & 0 \\[1em]
0 & 0 & 0 & 0 & 0 & 0 \\[1em]
\begin{aligned} m_w v(t_r \cos(\beta - \delta_3) \\ + r_r \sin(\beta - \delta_3)) \end{aligned} & \begin{aligned} &m_w(r_r^2 + t_r^2 \\ &+ ar_r \cos\delta_3 \\ &- l_r t_r \cos\delta_3 \\ &+ at_r \sin\delta_3 \\ &+ l_r r_r \sin\delta_3) \\ &+ J_w \end{aligned} & 0 & 0 & m_w(r_r^2 + t_r^2) + J_w & 0 \\[3em]
0 & 0 & 0 & 0 & 0 & 0
\end{bmatrix},
$$

$$(A.33)$$

$M_4 =$

$$
\begin{bmatrix}
m_w v^2 & \begin{array}{c} -m_w v(l_r \cos\beta \\ + a\sin\beta \\ - t_r \cos(\beta - \delta_4) \\ + r_r \sin(\beta - \delta_4)) \end{array} & 0 & 0 & 0 & \begin{array}{c} m_w v(t_r \cos(\beta - \delta_4) \\ - r_r \sin(\beta - \delta_4)) \end{array} \\[2em]
\begin{array}{c} -m_w v(l_r \cos\beta \\ + a\sin\beta \\ - t_r \cos(\beta - \delta_4) \\ + r_r \sin(\beta - \delta_4)) \end{array} & \begin{array}{c} m_w(a^2 + l_r^2 \\ + r_r^2 + t_r^2 \\ + 2ar_r \cos\delta_4 \\ - 2l_r t_r \cos\delta_4 \\ - 2at_r \sin\delta_4 \\ - 2l_r r_r \sin\delta_4) \\ + J_w \end{array} & 0 & 0 & 0 & \begin{array}{c} m_w(r_r^2 + t_r^2 \\ + ar_r \cos\delta_4 \\ - l_r t_r \cos\delta_4 \\ - at_r \sin\delta_4 \\ - l_r r_r \sin\delta_4) \\ + J_w \end{array} \\[2em]
0 & 0 & 0 & 0 & 0 & 0 \\[0.5em]
0 & 0 & 0 & 0 & 0 & 0 \\[0.5em]
0 & 0 & 0 & 0 & 0 & 0 \\[0.5em]
\begin{array}{c} m_w v(t_r \cos(\beta - \delta_4) \\ - r_r \sin(\beta - \delta_4)) \end{array} & \begin{array}{c} m_w(r_r^2 + t_r^2 \\ + ar_r \cos\delta_4 \\ - l_r t_r \cos\delta_4 \\ - at_r \sin\delta_4 \\ - l_r r_r \sin\delta_4) \\ + J_w \end{array} & 0 & 0 & 0 & m_w(r_r^2 + t_r^2) + J_w
\end{bmatrix}.
$$

$$\tag{A.34}$$

Components $k_i = L_{T_i}^T m_i \overline{a}_i + L_{R_i}^T I_i \overline{\alpha}_i + L_{R_i}^T (\omega_i \times I_i \omega_i)$ of generalised Coriolis and centrifugal forces (2.28) are

$$k_0 = \begin{bmatrix} \omega m_b v^2 & 0 & 0 & 0 & 0 & 0 \end{bmatrix}^T, \tag{A.35}$$

$$
k_1 = \begin{bmatrix}
\begin{array}{c} m_w v(\omega v - \omega^2(a\cos\beta - l_f \sin\beta)) \\ - (\dot{\delta}_1 + \omega)^2 (r_f \cos(\beta - \delta_1) - t_f \sin(\beta - \delta_1)) \end{array} \\[1.5em]
\begin{array}{c} m_w(\omega v(l_f \cos\beta + a\sin\beta + t_f \cos(\beta - \delta_1) + r_f \sin(\beta - \delta_1)) \\ + (\dot{\delta}_1^2 + 2\dot{\delta}_1 \omega)((at_f - l_f r_f)\cos\delta_1 - (ar_f + l_f t_f)\sin\delta_1)) \end{array} \\[1.5em]
\begin{array}{c} m_w(\omega v(t_f \cos(\beta - \delta_1) + r_f \sin(\beta - \delta_1)) \\ - \omega^2((at_f - l_f r_f)\cos\delta_1 - (ar_f + l_f t_f)\sin\delta_1)) \end{array} \\[1.5em]
0 \\[0.5em]
0 \\[0.5em]
0
\end{bmatrix},
$$

$$\tag{A.36}$$

$$
k_2 = \begin{bmatrix}
\begin{aligned}
& m_w v(\omega v + \omega^2(a\cos\beta + l_f\sin\beta)) \\
& + (\dot\delta_2 + \omega)^2(r_f\cos(\beta - \delta_2) + t_f\sin(\beta - \delta_2))
\end{aligned} \\[2mm]
\begin{aligned}
& m_w(\omega v(l_f\cos\beta - a\sin\beta + t_f\cos(\beta - \delta_2) - r_f\sin(\beta - \delta_2)) \\
& - (\dot\delta_2^2 + 2\dot\delta_2\omega)((at_f - l_f r_f)\cos\delta_2 + (ar_f + l_f t_f)\sin\delta_2))
\end{aligned} \\[2mm]
0 \\[2mm]
\begin{aligned}
& m_w(\omega v(t_f\cos(\beta - \delta_2) - r_f\sin(\beta - \delta_2)) \\
& + \omega^2((at_f - l_f r_f)\cos\delta_2 + (ar_f + l_f t_f)\sin\delta_2))
\end{aligned} \\[2mm]
0 \\[2mm]
0
\end{bmatrix},
$$

(A.37)

$$
k_3 = \begin{bmatrix}
\begin{aligned}
& m_w v(\omega v - \omega^2(a\cos\beta - l_r\sin\beta)) \\
& - (\dot\delta_1 + \omega)^2(r_r\cos(\beta - \delta_3) - t_r\sin(\beta - \delta_3))
\end{aligned} \\[2mm]
\begin{aligned}
& -m_w(\omega v(l_r\cos\beta - a\sin\beta - t_r\cos(\beta - \delta_3) - r_r\sin(\beta - \delta_3)) \\
& - (\dot\delta_3^2 + 2\dot\delta_3\omega)((at_r + l_r r_r)\cos\delta_3 - (ar_r + l_r t_r)\sin\delta_3))
\end{aligned} \\[2mm]
0 \\[2mm]
0 \\[2mm]
\begin{aligned}
& m_w(\omega v(t_r\cos(\beta - \delta_3) + r_r\sin(\beta - \delta_3)) \\
& - \omega^2((at_r + l_r r_r)\cos\delta_3 - (ar_r - l_r t_r)\sin\delta_3))
\end{aligned} \\[2mm]
0
\end{bmatrix},
$$

(A.38)

$$
k_4 = \begin{bmatrix}
\begin{aligned}
& m_w v(\omega v + \omega^2(a\cos\beta - l_r\sin\beta)) \\
& + (\dot\delta_4 + \omega)^2(r_r\cos(\beta - \delta_4) + t_r\sin(\beta - \delta_4))
\end{aligned} \\[2mm]
\begin{aligned}
& -m_w(\omega v(l_r\cos\beta + a\sin\beta - t_r\cos(\beta - \delta_4) + r_r\sin(\beta - \delta_4)) \\
& + (\dot\delta_4^2 + 2\dot\delta_4\omega)((at_r + l_r r_r)\cos\delta_4 + (ar_r - l_r t_r)\sin\delta_4))
\end{aligned} \\[2mm]
0 \\[2mm]
\begin{aligned}
& m_w(\omega v(t_r\cos(\beta - \delta_4) - r_r\sin(\beta - \delta_4)) \\
& + \omega^2((at_r + l_r r_r)\cos\delta_4 + (ar_r - l_r t_r)\sin\delta_4))
\end{aligned} \\[2mm]
0 \\[2mm]
0
\end{bmatrix}.
$$

(A.39)

Components $\boldsymbol{q}_i = \boldsymbol{L}_{T_i}^{\mathrm{T}} \boldsymbol{f}_i^a + \boldsymbol{L}_{R_i}^{\mathrm{T}} \boldsymbol{l}_i^a$ of generalised applied forces (2.29) are

$$\boldsymbol{q}_0 = \begin{bmatrix} 0 & c_f\left(\delta_1 + \delta_2\right) + c_r\left(\delta_3 + \delta_4\right) + d \sum_{i=1\ldots 4} \dot{\delta}_i & 0 & 0 & 0 & 0 \end{bmatrix}^{\mathrm{T}}, \quad (A.40)$$

$$\boldsymbol{q}_1 = \begin{bmatrix} F_{y_1} v \cos\left(\beta - \delta_1\right) - F_{x_1} v \sin\left(\beta - \delta_1\right) \\ \begin{aligned} & M_{z_1} - c_f\delta_1 - d\dot{\delta}_1 - F_{x_1}\left(r_f + a\cos\delta_1 - l_f\sin\delta_1\right) \\ & + F_{y_1}\left(t_f + l_f\cos\delta_1 + a\sin\delta_1\right) \end{aligned} \\ M_{z_1} - c_f\delta_1 - d\dot{\delta}_1 - F_{x_1}r_f + F_{y_1}t_f \\ 0 \\ 0 \\ 0 \end{bmatrix}, \quad (A.41)$$

$$\boldsymbol{q}_2 = \begin{bmatrix} F_{y_2} v \cos\left(\beta - \delta_2\right) - F_{x_2} v \sin\left(\beta - \delta_2\right) \\ \begin{aligned} & M_{z_2} - c_f\delta_2 - d\dot{\delta}_2 + F_{x_2}\left(r_f + a\cos\delta_2 + l_f\sin\delta_2\right) \\ & + F_{y_2}\left(t_f + l_f\cos\delta_2 - a\sin\delta_2\right) \end{aligned} \\ 0 \\ M_{z_2} - c_f\delta_2 - d\dot{\delta}_2 + F_{x_2}r_f + F_{y_2}t_f \\ 0 \\ 0 \end{bmatrix}, \quad (A.42)$$

$$\boldsymbol{q}_3 = \begin{bmatrix} F_{y_3} v \cos\left(\beta - \delta_3\right) - F_{x_3} v \sin\left(\beta - \delta_3\right) \\ \begin{aligned} & M_{z_3} - c_f\delta_3 - d\dot{\delta}_3 - F_{x_3}\left(r_r + a\cos\delta_3 + l_r\sin\delta_3\right) \\ & + F_{y_3}\left(t_r - l_r\cos\delta_3 + a\sin\delta_3\right) \end{aligned} \\ 0 \\ 0 \\ M_{z_3} - c_f\delta_3 - d\dot{\delta}_3 - F_{x_3}r_r + F_{y_3}t_r \\ 0 \end{bmatrix}, \quad (A.43)$$

$$\boldsymbol{q}_4 = \begin{bmatrix} F_{y_4} v \cos\left(\beta - \delta_4\right) - F_{x_4} v \sin\left(\beta - \delta_4\right) \\ \begin{aligned} & M_{z_4} - c_f\delta_4 - d\dot{\delta}_4 + F_{x_4}\left(r_r + a\cos\delta_4 - l_r\sin\delta_4\right) \\ & + F_{y_4}\left(t_r - l_r\cos\delta_4 - a\sin\delta_4\right) \end{aligned} \\ 0 \\ 0 \\ 0 \\ M_{z_4} - c_f\delta_4 - d\dot{\delta}_4 + F_{x_4}r_r + F_{y_4}t_r \end{bmatrix}. \quad (A.44)$$

A.3 Constraints

Position constraints (2.10) read in explicit form as

$$
c = \begin{bmatrix} (h_f \cos \delta_2 + c \sin \delta_2 - h_f \cos \delta_1 + c \sin \delta_1)^2 \\ + (2a + h_f \sin \delta_2 - c \cos \delta_2 - h_f \sin \delta_1 - c \cos \delta_1)^2 - b^2 \\ \\ (c \sin \delta_4 - h_r \cos \delta_4 + h_r \cos \delta_3 + c \sin \delta_3)^2 \\ + (2a - h_r \sin \delta_4 - c \cos \delta_4 + h_r \sin \delta_3 - c \cos \delta_3)^2 - b^2 \end{bmatrix} . \quad (A.45)
$$

The position constraint Jacobian results from differentiation of (A.45) w.r.t. (2.1) according to Eq. (2.22) as

$$
C_y = \frac{\partial c}{\partial y} = \begin{bmatrix} 0 & 0 & 0 & C_y^{(1,4)} & C_y^{(1,5)} & 0 & 0 \\ 0 & 0 & 0 & 0 & 0 & C_y^{(2,6)} & C_y^{(2,7)} \end{bmatrix}, \quad (A.46)
$$

where

$$
C_y^{(1,4)} = 2(c \cos \delta_1 + h_f \sin \delta_1)(h_f \cos \delta_2 - h_f \cos \delta_1 + c \sin \delta_1 + c \sin \delta_2) \\ - 2(c \sin \delta_1 - h_f \cos \delta_1)(c \cos \delta_1 + c \cos \delta_2 - h_f \sin \delta_2 + h_f \sin \delta_1 - 2a),
$$

$$
C_y^{(1,5)} = 2(c \cos \delta_2 - h_f \sin \delta_2)(h_f \cos \delta_2 - h_f \cos \delta_1 + c \sin \delta_1 + c \sin \delta_2) \\ - 2(c \sin \delta_2 + h_f \cos \delta_2)(c \cos \delta_1 + c \cos \delta_2 - h_f \sin \delta_2 + h_f \sin \delta_1 - 2a),
$$

$$
C_y^{(2,6)} = 2(c \cos \delta_3 - h_r \sin \delta_3)(h_r \cos \delta_3 - h_r \cos \delta_4 + c \sin \delta_3 + c \sin \delta_4) \\ - 2(h_r \cos \delta_3 + c \sin \delta_3)(c \cos \delta_3 + c \cos \delta_4 + h_r \sin \delta_4 - h_r \sin \delta_3 - 2a),
$$

$$
C_y^{(2,7)} = 2(c \cos \delta_4 + h_r \sin \delta_4)(h_r \cos \delta_3 - h_r \cos \delta_4 + c \sin \delta_3 + c \sin \delta_4) \\ + 2(h_r \cos \delta_4 - c \sin \delta_4)(c \cos \delta_3 + c \cos \delta_4 + h_r \sin \delta_4 - h_r \sin \delta_3 - 2a). \quad (A.47)
$$

The velocity constraint Jacobian (2.23) follows from (A.46) as

$$
C_z = C_y \frac{\partial f_v}{\partial z} = \begin{bmatrix} 0 & 0 & C_y^{(1,4)} & C_y^{(1,5)} & 0 & 0 \\ 0 & 0 & 0 & 0 & C_y^{(2,6)} & C_y^{(2,7)} \end{bmatrix}, \quad (A.48)
$$

where

$$
\frac{\partial f_v}{\partial z} = \begin{bmatrix} \begin{bmatrix} -v \sin (\beta + \psi) \\ v \cos (\beta + \psi) \\ 0^{5 \times 1} \end{bmatrix} & 0^{2 \times 5} \\ & I^{5 \times 5} \end{bmatrix} \quad (A.49)
$$

results from differentiation of (2.3) w.r.t. (2.2).

The remaining term of the acceleration constraints (2.34) reads as

$$
\gamma = \begin{bmatrix} 2(h_f^2 - c^2)(\dot{\delta}_1 - \dot{\delta}_2)^2\cos(\delta_1 - \delta_2) - 4ch_f(\dot{\delta}_1 - \dot{\delta}_2)^2\sin(\delta_1 - \delta_2) \\ - 4ah_f(\dot{\delta}_2^2\sin\delta_2 - \dot{\delta}_1^2\sin\delta_1) \\ \\ 2(h_r^2 - c^2)(\dot{\delta}_3 - \dot{\delta}_4)^2\cos(\delta_3 - \delta_4) + 4ch_r(\dot{\delta}_3 - \dot{\delta}_4)^2\sin(\delta_3 - \delta_4) \\ + 4ah_r(\dot{\delta}_4^2\sin\delta_4 - \dot{\delta}_3^2\sin\delta_3) \end{bmatrix}.
$$

$$(A.50)$$

List of Figures

List of Tables

References

ANDREWS, L. (1992). *Special Functions of Mathematics for Engineers.* SPIE Optical Engineering Press, Bellingham.

BAPST, R., JAKOB, M., ONDER, C., GUZZELLA, L., AND ASPRION, J. (2014). Model-linearization strategies for MPC of the air-path of a Diesel engine. In: *Proceedings of the 4th International Conference on Engineering Optimization (ENGOPT 2014)*, Lisbon, pp. 627–632.

BESTLE, D. (1994). *Analyse und Optimierung von Mehrkörpersystemen.* Springer, Berlin.

BLUNDELL, M. AND HARTY, D. (2015). *The Multibody Systems Approach to Vehicle Dynamics.* Butterworth-Heinemann, Oxford.

BODE, H. W. (1940). Relations Between Attenuation and Phase in Feedback Amplifier Design. *The Bell System Technical Journal*, 19(3), pp. 421–454.

BUSCH, J. (2015). *Unterstützende Strategien zur Optimierung der Fahrzeugquerdynamik.* Shaker, Aachen.

BUSCH, J. AND BESTLE, D. (2011). Optimization of a new steering strategy to improve the driving behavior of cars. In: *Proceedings of the 7th European Nonlinear Dynamics Conference (ENOC 2011)*, Rome, pp. 184–189.

BUSCH, J. AND BESTLE, D. (2014). Optimisation of lateral car dynamics taking into account parameter uncertainties. *Vehicle System Dynamics*, 52(2), pp. 166–185.

CAO, M., HU, C., WANG, J., WANG, R., AND CHEN, N. (2020). Adaptive complementary filter-based post-impact control for independently-actuated and differentially-steered autonomous vehicles. *Mechanical Systems and Signal Processing*, 144. Article No. 106852.

CHELI, F., SABBIONI, E., PESCE, M., AND MELZI, S. (2007). A methodology for vehicle sideslip angle identification: comparison with experimental data. *Vehicle System Dynamics*, 45(6), pp. 549–563.

CHEN, Y. AND WANG, J. (2012). Fast and Global Optimal Energy-Efficient Control Allocation With Applications to Over-Actuated Electric Ground Vehicles. *IEEE Transactions on Control Systems Technology*, 20(5), pp. 1202–1211.

CHEN, Y. AND WANG, J. (2014). Adaptive Energy-Efficient Control Allocation for Planar Motion Control of Over-Actuated Electric Ground Vehicles. *IEEE Transactions on Control Systems Technology*, 22(4), pp. 1362–1373.

CHO, J., KIM, K., JEON, D., AND YOO, W. (2005). Transient dynamic response analysis of 3-d patterned tire rolling over cleat. *European Journal of Mechanics - A/Solids*, 24(3), pp. 519–531.

CRESSIE, N. A. C. (1993). *Statistics for Spatial Data*. Wiley, New York, pp. 105–210.

DAN FORESEE, F. AND HAGAN, M. T. (1997). Gauss-Newton approximation to Bayesian learning. In: *Proceedings of the International Conference on Neural Networks (ICNN'97)*, Houston, pp. 1930–1935.

DEB, K. (2001). *Multi-Objective Optimization Using Evolutionary Algorithms*. Wiley, New York.

DELASALLE, A. (1899). Voiture électrique Vedovelli et Priestley. *La Locomotion Automobile*, 6(33), pp. 527–530.

DOMINGUEZ-GARCIA, A. D., KASSAKIAN, J. G., AND SCHINDALL, J. E. (2004). A backup system for automotive steer-by-wire, actuated by selective braking. In: *Proceedings of the 2004 IEEE 35th Annual Power Electronics Specialists Conference (IEEE Cat. No.04CH37551)*, Aachen, pp. 383–388.

DORNHEGE, J., NOLDEN, S., AND MAYER, M. (2017). Steering Torque Disturbance Rejection. *SAE International Journal of Vehicle Dynamics, Stability, and NVH*, 1(2), pp. 165–172.

ENGELMANN, D. AND HERR, S. (2018). Validation of a Mechanical Drive with Individual Wheel Control. *ATZ offhighway worldwide*, 11(1), pp. 8–15.

GAUGER, A., KERN, A., FEINAUER, J., KANNGIESSER, S., AND GREUL, R. (2016). Potential of wheel-individual brake interventions as a backup for steering system failures during automated driving. In: *Proceedings of the 7th International Munich Chassis Symposium*, Munich, pp. 485–503.

GE, X.-S., ZHAO, W.-J., CHEN, L.-Q., AND LIU, Y.-Z. (2005). Symbolic Linearization of Differential/Algebraic Equations Based on Cartesian Coordinates. *Technische Mechanik*, 25(3-4), pp. 230–240.

GONZÁLEZ, F., MASARATI, P., CUADRADO, J., AND NAYA, M. A. (2017). Assessment of Linearization Approaches for Multibody Dynamics Formulations. *Journal of Computational and Nonlinear Dynamics*, 12(4). Article No. 041009.

HAND, L. N. AND FINCH, J. D. (1998). *Analytical Mechanics*. Cambridge University Press, Cambridge, pp. 51–62.

HIRSCHBERG, W., RILL, G., AND WEINFURTER, H. (2007). Tire model TMeasy. *Vehicle System Dynamics*, 45(sup1), pp. 101–119.

HARRER, M. and PFEFFER, P., editors (2017). *Steering Handbook*. Springer, Cham.

HU, C., QIN, Y., CAO, H., SONG, X., JIANG, K., RATH, J. J., AND WEI, C. (2019). Lane keeping of autonomous vehicles based on differential steering with adaptive multivariable super-twisting control. *Mechanical Systems and Signal Processing*, 125, pp. 330–346.

HWANG, R. S., BAE, D. S., KUHL, J. G., AND HAUG, E. J. (1990). Parallel Processing for Real-Time Dynamic System Simulation. *Journal of Mechanical Design*, 112(4), pp. 520–528.

JONASSON, M. AND THOR, M. (2018). Steering redundancy for self-driving vehicles using differential braking. *Vehicle System Dynamics*, 56(5), pp. 791–809.

JOURDAIN, P. E. B. (1909). Note on an analogue of Gauss' principle of least constraint. *The Quarterly Journal of Pure and Applied Mathematics*, 40, pp. 153–157.

KAMEN, D. L., AMBROGI, R. R., DUGGAN, R. J., HEINZMANN, R. K., KEY, B. R., SKOSKIEWICZ, A., AND KRISTAL, P. K. (1994). Human transporter. Patent No. US5701965A.

KIRLI, A., OKWUDIRE, C. E., AND ULSOY, A. G. (2017). Limitations of Torque Vectoring as a Backup Safety Strategy for Steer-by-Wire Vehicles due to Vehicle Stability Control. In: *Proceedings of the ASME 2017 Dynamic Systems and Control Conference.*

KUSLITS, M. AND BESTLE, D. (2018a). Control design and performance evaluation of a new steering concept based on a custom multibody vehicle model. In: *Proceedings of the 5th International Conference on Dynamic Simulation in Vehicle Engineering*, Steyr, pp. 19–34.

KUSLITS, M. AND BESTLE, D. (2018b). Symbolic Linearized Equations for Nonholonomic Multibody Systems with Closed-Loop Kinematics. In: *Proceedings of the ASME 2018 International Design Engineering Technical Conferences and Computers and Information in Engineering Conference*, Quebec City. Article No. DETC2018-85823, V006T09A003.

KUSLITS, M. AND BESTLE, D. (2019). Modelling and control of a new differential steering concept. *Vehicle System Dynamics*, 57(4), pp. 520–542.

KUSLITS, M. AND BESTLE, D. (2022). Multiobjective performance optimisation of a new differential steering concept. *Vehicle System Dynamics*, 60(1), pp. 73–95.

LAULUSA, A. AND BAUCHAU, O. A. (2008). Review of Classical Approaches for Constraint Enforcement in Multibody Systems. *Journal of Computational and Nonlinear Dynamics*, 3(1). Article No. 011004.

LESHNO, M., LIN, V. Y., PINKUS, A., AND SCHOCKEN, S. (1993). Multilayer Feedforward Networks With a Nonpolynomial Activation Function Can Approximate Any Function. *Neural Networks*, 6(6), pp. 861–867.

LEVINE, W. (2011a). *Control System Advanced Methods.* CRC Press, Boca Raton.

LEVINE, W. (2011b). *Control System Fundamentals.* CRC Press, Boca Raton.

LI, Q., YU, X., ZHANG, H., AND HUANG, R. (2015). Study on Differential Assist Steering System with Double In-Wheel Motors with Intelligent Controller. *Mathematical Problems in Engineering*, 2015. Article No. 910230.

LI, S., YANG, J., CHEN, W., AND CHEN, X. (2014). *Disturbance Observer-Based Control: Methods and Applications.* CRC Press, Boca Raton.

Lundahl, K., Åslund, J., and Nielsen, L. (2011). Investigating Vehicle Model Detail for Close to Limit Maneuvers Aiming at Optimal Control. In: *Proceedings of the 22nd IAVSD Symposium on Dynamics of Vehicles on Roads and Tracks*, Manchester.

Maclaurin, B. (2008). Comparing the steering performances of skid- and Ackermann-steered vehicles. *Proceedings of the Institution of Mechanical Engineers, Part D: Journal of Automobile Engineering*, 222(5), pp. 739–756.

Matta, C. F., Massa, L., Gubskaya, A. V., and Knoll, E. (2011). Can One Take the Logarithm or the Sine of a Dimensioned Quantity or a Unit? Dimensional Analysis Involving Transcendental Functions. *Journal of Chemical Education*, 88(1), pp. 67–70.

McGuigan, S. J. and Moss, P. J. (1998). A Review of Transmission Systems for Tracked Military Vehicles. *Journal of Battlefield Technology*, 1(3), pp. 7–14.

Morris, M. D. and Mitchell, T. J. (1995). Exploratory designs for computational experiments. *Journal of Statistical Planning and Inference*, 43(3), pp. 381–402.

Motoyama, S., Uki, H., Isoda, K., and Yuasa, H. (1993). Effect of traction force distribution control on vehicle dynamics. *Vehicle System Dynamics*, 22(5-6), pp. 455–464.

Mousseau, C. and Hulbert, G. (1996). An efficient tire model for the analysis of spindle forces produced by a tire impacting large obstacles. *Computer Methods in Applied Mechanics and Engineering*, 135(1), pp. 15–34.

Nah, J., Yun, S., Yi, K., Kim, W., and Kim, J. (2013). Driving Control Architecture for Six-In-Wheel-Driving and Skid-Steered Series Hybrid Vehicles. In: *Proceedings of the 2013 World Electric Vehicle Symposium and Exhibition (EVS27)*, pp. 915–921. IEEE.

Negrut, D. and Ortiz, J. L. (2006). A Practical Approach for the Linearization of the Constrained Multibody Dynamics Equations. *Journal of Computational and Nonlinear Dynamics*, 1(3), pp. 230–239.

Neuman, C. P. and Murray, J. J. (1984). Linearization and Sensitivity Functions of Dynamic Robot Models. *IEEE Transactions on Systems, Man, and Cybernetics*, SMC-14(6), pp. 805–818.

NGUYEN, T.-A. AND BESTLE, D. (2007). Application of Optimization Methods to Controller Design for Active Suspensions. *Mechanics Based Design of Structures and Machines*, 35(3), pp. 291–318.

OGATA, K. (2010). *Modern Control Engineering*. Prentice Hall, Upper Saddle River.

OKE, P. AND NGUANG, S. K. (2020). Robust H$_\infty$ Takagi–Sugeno fuzzy output-feedback control for differential speed steering vehicles. *Proceedings of the Institution of Mechanical Engineers, Part D: Journal of Automobile Engineering*, 234(12), pp. 2822–2835.

OPPENHEIMER, M. W., DOMAN, D. B., AND BOLENDER, M. A. (2006). Control Allocation for Over-actuated Systems. In: *Proceedings of the 2006 14th Mediterranean Conference on Control and Automation*, Ancona, pp. 819–824.

PACEJKA, H. B. (2006). *Tyre and Vehicle Dynamics*. Butterworth-Heinemann, Oxford.

PAUWELUSSEN, J. P. (2014). *Essentials of Vehicle Dynamics*. Butterworth-Heinemann, Oxford.

PEROVIC, D. K. (2012). Making the Impossible, Possible – Overcoming the Design Challenges of In Wheel Motors. *World Electric Vehicle Journal*, 5(2), pp. 514–519.

PETERSON, D. L., GEDE, G., AND HUBBARD, M. (2015). Symbolic linearization of equations of motion of constrained multibody systems. *Multibody System Dynamics*, 33(2), pp. 143–161.

POLMANS, K. AND STRACKE, S. (2014). Torque vectoring as redundant steering for automated driving or steer-by-wire. In: *Proceedings of the 5th International Munich Chassis Symposium*, Munich, pp. 163–177.

POWELL, M. J. D. (1987). Radial Basis Functions for Multivariable Interpolation: A Review. In: Mason, J. C. and Cox, M. G., editors, *Algorithms for Approximation*, pp. 143–167. Oxford University Press, Oxford.

REGIS, R. G. AND SHOEMAKER, C. A. (2004). Local Function Approximation in Evolutionary Algorithms for the Optimization of Costly Functions. *IEEE Transactions on Evolutionary Computation*, 8(5), pp. 490–505.

REITER, G., POLMANS, K., MIANO, C., HACKL, A., AND LEX, C. (2018). Suspension influences on a steer-by-wire torque vectoring vehicle. In: *Proceedings of the 18. Internationales Stuttgarter Symposium*, Stuttgart, pp. 327–341.

RIEKERT, P. AND SCHUNCK, T. E. (1940). Zur Fahrmechanik des gummibereiften Kraftfahrzeugs. *Ingenieur-Archiv*, 11(3), pp. 210–224.

RILL, G. (2012). *Road Vehicle Dynamics*. CRC Press, Boca Raton.

RÖMER, J., KAUTZMANN, P., FREY, M., AND GAUTERIN, F. (2018). Reducing Energy Demand Using Wheel-Individual Electric Drives to Substitute EPS-Systems. *Energies*, 11(1). Article No. 247.

SAKAI, S., SADO, H., AND HORI, Y. (1999). Motion Control in an Electric Vehicle with Four Independently Driven In-Wheel Motors. *IEEE/ASME Transactions on Mechatronics*, 4(1), pp. 9–16.

SHAMMA, J. S. AND ATHANS, M. (1991). Guaranteed Properties of Gain Scheduled Control for Linear Parameter-varying Plants. *Automatica*, 27(3), pp. 559–564.

STAICU, S. (2009). Dynamics equations of a mobile robot provided with caster wheel. *Nonlinear Dynamics*, 58(1), pp. 237–248.

TIAN, J., WANG, Q., DING, J., WANG, Y., AND MA, Z. (2019). Integrated Control With DYC and DSS for 4WID Electric Vehicles. *IEEE Access*, 7, pp. 124077–124086.

TULPULE, P. (2014). *Integrated Robust Optimal Design (IROD) via sensitivity minimization*. PhD thesis, Iowa State University.

VENTER, G., HAFTKA, R., AND JAMES STARNES, J. (1996). Construction of response surfaces for design optimization applications. In: *Proceedings of the 6th Symposium on Multidisciplinary Analysis and Optimization*, Bellevue, pp. 548–564.

WADEPHUL, J., ENGELMANN, D., KAUTZMANN, P., AND RÖMER, J. (2018). Development of a Novel Steering Concept for Articulated Vehicles. *ATZ offhighway worldwide*, 11(4), pp. 54–61.

WANG, H., KONG, H., MAN, Z., TUAN, D. M., CAO, Z., AND SHEN, W. (2014). Sliding Mode Control for Steer-by-Wire Systems With AC Motors in Road Vehicles. *IEEE Transactions on Industrial Electronics*, 61(3), pp. 1596–1611.

WANG, J., WANG, Q., AND JIN, L. (2008). Modeling and Simulation Studies on Differential Drive Assisted Steering for EV with Four-Wheel-Independent-Drive. In: *Proceedings of the 2008 IEEE Vehicle Power and Propulsion Conference (VPPC '08)*, Harbin, pp. 682–689.

WANG, J., WANG, Q., JIN, L., AND SONG, C. (2011). Independent wheel torque control of 4WD electric vehicle for differential drive assisted steering. *Mechatronics*, 21(1), pp. 63–76.

WANG, J., WANG, X., LUO, Z., AND ASSADIAN, F. (2020a). Active Disturbance Rejection Control of Differential Drive Assist Steering for Electric Vehicles. *Energies*, 13(10). Article No. 2647.

WANG, J., YAN, T., BAI, Y., LUO, Z., LI, X., AND YANG, B. (2020b). Assistance Quality Analysis and Robust Control of Electric Vehicle With Differential Drive Assisted Steering System. *IEEE Access*, 8, pp. 136327–136339.

WANG, Q., WANG, J., AND JIN, L. (2009). Driver-Vehicle Closed-Loop Simulation of Differential Drive Assist Steering Control System for Motorized-Wheel Electric Vehicle. In: *Proceedings of the 5th IEEE Vehicle Power and Propulsion Conference (VPPC '09)*, Dearborn, pp. 564–571.

WANG, R. AND WANG, J. (2011). Fault-Tolerant Control With Active Fault Diagnosis for Four-Wheel Independently Driven Electric Ground Vehicles. *IEEE Transactions on Vehicular Technology*, 60(9), pp. 4276–4287.

WEHBI, K., WURM, A., BESTLE, D., AND BEILHARZ, J. (2017). Automated Calibration Using Simulation and Robust Design Optimization Improving Shift and Launch Quality of Automatic Transmissions. In: *Proceedings of the Automotive data analytics, methods, DoE, proceedings of the International Calibration Conference*, Berlin, pp. 18–33.

WU, F., YEH, T., AND HUANG, C. (2013). Motor control and torque coordination of an electric vehicle actuated by two in-wheel motors. *Mechatronics*, 23(1), pp. 46–60.

ZHAO, W., LI, Y., WANG, C., ZHANG, Z., AND XU, C. (2013). Research on control strategy for differential steering system based on H mixed sensitivity. *International Journal of Automotive Technology*, 14(6), pp. 913–919.

ZHAO, W., WANG, C., SUN, P., AND LIU, S. (2011). Primary studies on integration optimization of differential steering of electric vehicle with motorized wheels based on quality engineering. *Science China Technological Sciences*, 54(11), pp. 3047–3053.

ZHAO, W., XU, X., AND WANG, C. (2012). Multidiscipline collaborative optimization of differential steering system of electric vehicle with motorized wheels. *Science China Technological Sciences*, 55(12), pp. 3462–3468.

ZHAO, W., YANG, Z., AND WANG, C. (2018). Multidisciplinary hybrid hierarchical collaborative optimization of electric wheel vehicle chassis integrated system based on driver's feel. *Structural and Multidisciplinary Optimization*, 57(3), pp. 1129–1147.

ZHAO, W., YANG, Z., AND WANG, C. (2019). Multi-objective optimization of chassis integrated system for electric wheel vehicle. *Proceedings of the Institution of Mechanical Engineers, Part C: Journal of Mechanical Engineering Science*, 233(1), pp. 7–17.

ZHAO, W. AND ZHANG, H. (2018). Coupling Control Strategy of Force and Displacement for Electric Differential Power Steering System of Electric Vehicle With Motorized Wheels. *IEEE Transactions on Vehicular Technology*, 67(9), pp. 8118–8128.